T0122207

Springer Theses

Recognizing Outstanding Ph.D. Research

Aims and Scope

The series "Springer Theses" brings together a selection of the very best Ph.D. theses from around the world and across the physical sciences. Nominated and endorsed by two recognized specialists, each published volume has been selected for its scientific excellence and the high impact of its contents for the pertinent field of research. For greater accessibility to non-specialists, the published versions include an extended introduction, as well as a foreword by the student's supervisor explaining the special relevance of the work for the field. As a whole, the series will provide a valuable resource both for newcomers to the research fields described, and for other scientists seeking detailed background information on special questions. Finally, it provides an accredited documentation of the valuable contributions made by today's younger generation of scientists.

Theses are accepted into the series by invited nomination only and must fulfill all of the following criteria

- They must be written in good English.
- The topic should fall within the confines of Chemistry, Physics, Earth Sciences, Engineering and related interdisciplinary fields such as Materials, Nanoscience, Chemical Engineering, Complex Systems and Biophysics.
- The work reported in the thesis must represent a significant scientific advance.
- If the thesis includes previously published material, permission to reproduce this must be gained from the respective copyright holder.
- They must have been examined and passed during the 12 months prior to nomination.
- Each thesis should include a foreword by the supervisor outlining the significance of its content.
- The theses should have a clearly defined structure including an introduction accessible to scientists not expert in that particular field.

More information about this series at http://www.springer.com/series/8790

Oliver Allanson

Theory of One-Dimensional Vlasov-Maxwell Equilibria

With Applications to Collisionless Current Sheets and Flux Tubes

Doctoral Thesis accepted by
the University of St Andrews, St Andrews, UK

Author
Dr. Oliver Allanson
Department of Meteorology
University of Reading
Reading, UK

Supervisor
Prof. Thomas Neukirch
University of St Andrews
St Andrews, UK

ISSN 2190-5053 ISSN 2190-5061 (electronic)
Springer Theses
ISBN 978-3-030-07372-5 ISBN 978-3-319-97541-2 (eBook)
https://doi.org/10.1007/978-3-319-97541-2

Softcover re-print of the Hardcover 1st edition 2018

This Springer imprint is published by the registered company Springer Nature Switzerland AG
The registered company address is: Gewerbestrasse 11, 6330 Cham, Switzerland

This thesis is dedicated to
Sophie Dawe's love, and levity
Phil Michaels' life, and spirit
my parents' support, and tender care
my family present, passed, and in-law
my friends for bringing me to life,
and cutting me down to size.

At quite uncertain times and places,
The atoms left their heavenly path,
And by fortuitous embraces,
Engendered all that being hath.
And though they seem to cling together,
And form "associations" here,
Yet, soon or late, they burst their tether,
And through the depths of space career.
Soon, all too soon, the chilly morning,
This flow of soul will crystallize,
Then those who Nonsense now are scorning,
May learn, too late, where wisdom lies.

James Clerk Maxwell
Molecular evolution (abridged)
Nature, **8**, 205, p. 473 (1873)

Supervisor's Foreword

Whilst plasmas, i.e. partially or fully ionised gases, have to be artificially generated on Earth (e.g. in fusion devices), they occur naturally in space and a large proportion of ordinary matter in the Universe is assumed to be in the plasma state. Within our own solar system, we can study plasmas across a huge range of parameter regimes. Some examples are the extremely dense, high-temperature and fully ionised plasma in the Sun's core, the more tenuous outer parts of the solar atmosphere (solar corona and solar wind) and planetary magnetospheres, in which fully ionised regions can be coupled to only partially ionised domains, for example Earth's ionosphere.

In more tenuous plasmas, collisions between individual particles can be neglected and only the collective interaction between the charged particles via electromagnetic forces has to be considered. These collisionless plasmas are theoretically described by the Vlasov (collisionless Boltzmann) equation combined with Maxwell's equations (for short called the Vlasov-Maxwell equations). This nonlinear system of equations has been used for many years to study plasma phenomena such as waves and instabilities, and the theory has been used to understand and explain activity processes in space and other plasmas.

As in many areas of physics, the basic time-independent (equilibrium) solutions of the Vlasov-Maxwell equations are important as a foundation for studies of time-dependent phenomena. Using the mathematical method of characteristics equilibrium solutions, the Vlasov equation can in principle be found; in practice, however, this is only possible if the plasma system has spatial symmetries that are associated with constants of motion. One such solution found is E. Harris in the early 1960s has been used in a large number of investigations.

The method for finding Vlasov-Maxwell equilibria described above starts with finding the particle distribution function first and then solves Maxwell's equations in a second step to obtain the full solution. In practice, however, one often wants to carry out these steps in reverse order, i.e. start with given electromagnetic fields and find consistent particle distribution functions that solve the Vlasov equation. This turns out to be a formidable problem, beset with mathematical difficulties.

Building on previous work, Oliver Allanson has in his Ph.D. thesis advanced this particular subject substantially. Underpinned by rigorous mathematical methods, he has for the first time found an analytical solution for a distribution function that is consistent with a magnetic field configuration, that is of particular interest for space and astrophysical plasmas—a nonlinear force-free magnetic field. The methodology developed in his thesis does, however, have much wider applications. Whereas the case just described is given in Cartesian geometry, the thesis also investigates whether solutions with similar properties can be found in cylindrical coordinates, which is another geometry that is potentially important for astrophysical plasmas. Finally, the thesis presents a new family of analytical distribution functions for asymmetric current sheets, which have applications, for example, to the interface between the terrestrial magnetosphere and the solar wind, the magnetopause.

In my opinion, the value of this thesis does not only lie in its contents, but also in the potential for future work based on the methodology developed in it.

St Andrews, UK
June 2018

Prof. Thomas Neukirch

Abstract

Vlasov-Maxwell equilibria are characterised by the self-consistent descriptions of the steady states of collisionless plasmas in particle phase space and balanced macroscopic forces. We study the theory of Vlasov-Maxwell equilibria in one spatial dimension, as well as its application to current sheet and flux tube models.

The 'inverse problem' is that of determining a Vlasov-Maxwell equilibrium distribution function self-consistent with a given magnetic field. We develop the theory of inversion using expansions in Hermite polynomial functions of the canonical momenta. Sufficient conditions for the convergence of a Hermite expansion are found, given a pressure tensor. For large classes of DFs, we prove that non-negativity of the distribution function is contingent on the magnetisation of the plasma, and make conjectures for all classes.

The inverse problem is considered for nonlinear 'force-free Harris sheets'. By applying the Hermite method, we construct new models that can describe sub-unity values of the plasma beta (β_{pl}) for the first time. Whilst analytical convergence is proven for all β_{pl}, numerical convergence is attained for $\beta_{pl} = 0.85$, and then $\beta_{pl} = 0.05$ after a 're-gauging' process.

We consider the properties that a pressure tensor must satisfy to be consistent with 'asymmetric Harris sheets', and construct new examples. It is possible to analytically solve the inverse problem in some cases, but others must be tackled numerically. We present new exact Vlasov-Maxwell equilibria for asymmetric current sheets, which can be written as a sum of shifted Maxwellian distributions. This is ideal for implementations in particle-in-cell simulations.

We study the correspondence between the microscopic and macroscopic descriptions of equilibrium in cylindrical geometry and then attempt to find Vlasov-Maxwell equilibria for the nonlinear force-free 'Gold–Hoyle' model. However, it is necessary to include a background field, which can be arbitrarily weak if desired. The equilibrium can be electrically non-neutral, depending on the bulk flows.

Publications Related to this Thesis

The following published papers draw upon research conducted for this thesis:

1. O. Allanson, T. Neukirch, F. Wilson & S. Troscheit: An exact collisionless equilibrium for the Force-Free Harris Sheet with low plasma beta *Physics of Plasmas*, **22**, 102116, 2015.
2. O. Allanson, T. Neukirch, S. Troscheit & F. Wilson: From one-dimensional fields to Vlasov equilibria: theory and application of Hermite polynomials *Journal of Plasma Physics*, **82**, 905820306, 2016.
3. O. Allanson, F. Wilson & T. Neukirch: Neutral and non-neutral collisionless plasma equilibria for twisted flux tubes: The Gold-Hoyle model in a background field *Physics of Plasmas* **23**, 092106, 2016.
4. O. Allanson, F. Wilson, T. Neukirch, Y.-H. Liu, & J. D. B. Hodgson: Exact Vlasov-Maxwell equilibria for asymmetric current sheets *Geophysical Research Letters*, **44**, 2017.
5. O. Allanson, T. Neukirch and S. Troscheit: The inverse problem for Channell collisionless plasma equilibria *IMA Journal of Applied Mathematics*, hxy026 2018, (Invited paper).

Contents

Abbreviations

1D	One-dimensional
2D	Two-dimensional
AH+G	Asymmetric Harris plus guide
AHS	Asymmetric Harris sheet
DF	Distribution function
FFHS	Force-free Harris sheet
FT	Fourier transform
GEM	Geospace Environmental Modelling
GH	Gold–Hoyle
GH+B	Gold–Hoyle plus background
IFT	Inverse Fourier transform
LHS	Light-hand side
MHD	Magnetohydrodynamics
MMS	Magnetospheric MultiScale mission
PIC	Particle-in-cell
RHS	Right-hand side
VM	Vlasov-Maxwell

Some Important Notations

$\boldsymbol{x}$	(Particle) position
$\boldsymbol{v}$	(Particle) velocity, $\boldsymbol{v} = d\boldsymbol{x}/dt$
$\boldsymbol{A}$	Magnetic vector potential
$\boldsymbol{B}$	Magnetic field, $\boldsymbol{B} = \nabla \times \boldsymbol{A}$
ϕ	Electrostatic scalar potential
$\boldsymbol{E}$	Electric field, $\boldsymbol{E} = -\nabla\phi - \partial \mathbf{A}/\partial t$
$\alpha(\boldsymbol{r})$	Force-free parameter, $\alpha(\boldsymbol{r}) = \boldsymbol{B} \cdot (\nabla \times \boldsymbol{B})/\lvert\boldsymbol{B}\rvert^2$
s	Particle species s
m_s	Particle mass
q_s	Particle charge, $e = q_i = -q_e$
f_s	Particle distribution function (DF)
n_s	Particle number density, $n_s = \int f_s d^3v$
ρ_s	Mass density, $\rho_s = m_s n_s$
σ	Electric charge density, $\sigma = \sum_s \sigma_s = \sum_s q_s n_s$
$\boldsymbol{V}_s$	Bulk flow, $\boldsymbol{V}_s = n_s^{-1} \int \boldsymbol{v} f_s d^3v$
$\boldsymbol{j}$	Electric current density, $\boldsymbol{j} = \sum_s \boldsymbol{j}_s = \sum_s q_s n_s \boldsymbol{V}_s$
$\boldsymbol{w}_s$	Particle flow relative to the bulk, $\boldsymbol{w}_s = \boldsymbol{v} - \boldsymbol{V}_s$
P_{ij}	Thermal pressure tensor, $P_{ij} = \sum_s P_{ij,s} = \sum_s \int w_{is} w_{js} f_s d^3v$
p	Scalar thermal pressure, $p = \mathrm{Tr}(P_{ij})/3$
H_s	Particle Hamiltonian (energy), $H_s = m_s \boldsymbol{v}^2/2 + q_s\phi$
$\boldsymbol{p}_s$	Particle canonical momenta, $\boldsymbol{p}_s = m_s\boldsymbol{v} + q_s\boldsymbol{A}$
β_s	Thermal beta, $\beta_s = 1/(m_s v_{th,s}^2)$
$v_{\mathrm{th},s}$	Particle thermal velocity
r_{Ls}	Thermal Larmor radius, $r_{Ls} = m_s v_{th,s}(e\lvert\boldsymbol{B}\rvert)^{-1}$
T_s	Temperature, $T_s = 1/(k_B\beta_s)$

L	Macroscopic length scale (e.g. current sheet width)
δ_s	Magnetisation parameter, $\delta_s = r_{Ls}/L$
β_{pl}	Plasma beta, $\beta_{pl} = \sum_s \beta_{pl,s} = \sum_s n_s k_B T_s / (\boldsymbol{B}^2/(2/\mu_0))$
λ_D	Debye radius, $\lambda_D = \sqrt{\varepsilon_0 k_B T_e/(n_e e^2)}$

Physical Constants (SI Units)

Boltzmann's constant (k_B)	1.3807×10^{-23} J K^{-1}
Speed of light in a vacuum (c)	2.9979×10^{8} m s^{-1}
Permittivity of free space (ϵ_0)	8.8542×10^{-12} F m^{-1}
Permeability of free space (μ_0)	$4\pi \times 10^{-7}$ H m^{-1}
Proton mass (m_i)	1.6726×10^{-27} kg
Electron mass (m_e)	9.1094×10^{-31} kg
Elementary charge (e)	1.6022×10^{-19} C

Chapter 1
Introduction

Most important part of doing physics is the knowledge of approximation.

Lev Landau

1.1 The Hierarchy of Plasma Models

More than 99% of the known matter in the Universe is in the plasma state (Baumjohann and Treumann 1997), by far the most significant material constituent of stellar, interplanetary, interstellar and intergalactic media. Not only is a deep understanding of plasmas then clearly necessary to understand the physics of our universe, but plasmas are also of real interest to us on Earth. Nuclear fusion experiments—and in principle, future power stations—necessarily exploit the plasma state to work, either using high-temperature plasmas confined by strong magnetic fields, or plasmas formed by the laser ablation of a solid fuel target.

Plasmas are often known as the 'fourth' state of matter, lying after the 'third', and more familiar gaseous state. At a temperature above 100,000 K, most matter exists in an ionised state, however plasmas can exist at much lower temperatures should ionisation mechanisms exist, and if the density is sufficiently low (Krall and Trivelpiece 1973). Figures 1.1a, b display some examples from the rich array of plasma environments in temperature-density scatter plots; from the relatively cool and diffuse plasmas of interstellar space, to the incredibly dense and hot plasmas of stellar and laboratory fusion. Since there is such variety in the physical conditions able to sustain plasmas, the 'plasma state' may best describe *collective behaviours*, the characteristics that persist despite the range of physical conditions that can sustain plasmas (we see from Fig. 1.1b that even the free electrons in metals can be considered, or modelled, as a plasma). Matter is in a plasma state when the degree of ionisation is sufficiently high that the dynamical behaviour of the particles is dominated by electromagnetic forces (Fitzpatrick 2014), and this can even be the case for ionisation levels as low as a fraction of a percent (Peratt 1996). Whilst many of

O. Allanson, *Theory of One-Dimensional Vlasov-Maxwell Equilibria*,
Springer Theses, https://doi.org/10.1007/978-3-319-97541-2_1

Fig. 1.1 The variety of plasma conditions and environments

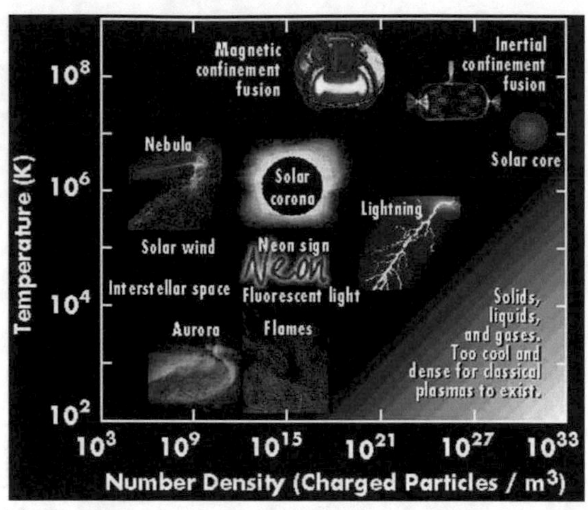

(a) The plasma 'zoo': A density-temperature plot displaying various plasma environments and phenomena, and their contrast to solids, liquids and gases. **Image copyright:** Contemporary Physics Education Project, (reproduced with permission).

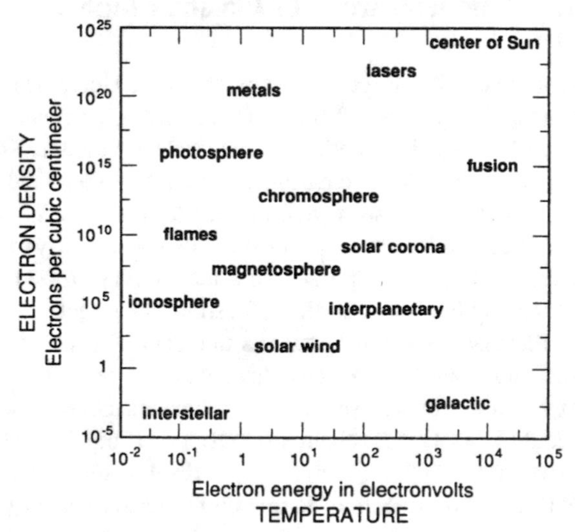

(b) A temperature-density plot reproduced from Peratt, (1996), focussing on the environments in which plasmas appear. **Image Copyright:** Springer, *Astrophysics and Space Science* **242**, 1-2, (1996), pp. 93-163., copyright (1996), (reproduced with permission).

these plasmas possess some shared tendencies and behaviours, it is not possible to capture all the detailed physics of the entire variety of plasma processes with one particular mathematical toolkit or model. Not only may some models fail to capture certain aspects of the physics by virtue of the approximations made, but they may be inefficient, or in fact insoluble when applied in practice. Hence, plasma physics is a discipline with a rich variety of perspectives and methods. Within each of these paradigms we make certain approximations and ordering assumptions, in order to capture the essence of the problem at hand.

1.1.1 Single Particle Motion

Taking the viewpoint of particulate matter as the fundamental approach, then a 'full' description of plasmas is found by solving the (Lorentz) equation of motion of each individual particle, written in classical form as

$$\boldsymbol{F}_s(\boldsymbol{x}(t), \boldsymbol{v}(t); t) = q_s(\boldsymbol{E}(\boldsymbol{x}, t) + \boldsymbol{v}(t) \times \boldsymbol{B}(\boldsymbol{x}, t)), \tag{1.1}$$

with the force, $\boldsymbol{F}_s$, on a test particle of species s, of charge q_s, at position $\boldsymbol{x}$, and with velocity $\boldsymbol{v}$, when under the influence of electric and magnetic fields, $\boldsymbol{E}$ and $\boldsymbol{B}$. One can in principle integrate in time to calculate the trajectory of the particle for all future times (e.g. see Vekstein et al. 2002),

$$\boldsymbol{x}(t) = \int_{t_0}^{t} \boldsymbol{v}(t')dt',$$

for $\boldsymbol{v}(t_0)$ some initial condition. However, in all but the simplest electromagnetic field geometries these integrals may not even be able to be written down, and/or one might have to resort to numerical methods to calculate the trajectory. One more complication is the effect of the charged particles on the electromagnetic fields, $\boldsymbol{E}$ and $\boldsymbol{B}$, and this shall be discussed in Sect. 1.1.2.

If a plasma is sufficiently magnetised it has small parameters

$$\frac{r_L}{L} \ll 1, \quad \frac{1/\Omega}{\tau} \ll 1,$$

for r_L and Ω the characteristic values of the Larmor radius and gyrofrequency of individual particle gyromotion respectively, and L and τ the characteristic length and time scales upon which the electromagnetic fields vary. In such a case there is a well understood treatment for particle orbits, namely *Guiding Centre* theory (e.g. see Northrop 1961; Littlejohn 1983; Cary and Brizard 2009). Guiding centre theory models particle motion as a superposition of rapid gyromotion and a comparitively slow *secular* drift (e.g. see Morozov and Solov'ev 1966). This gyromotion is depicted in Fig. 1.2, reproduced from Northrop (1963); in which the notation $\boldsymbol{\rho}$ and ρ are used

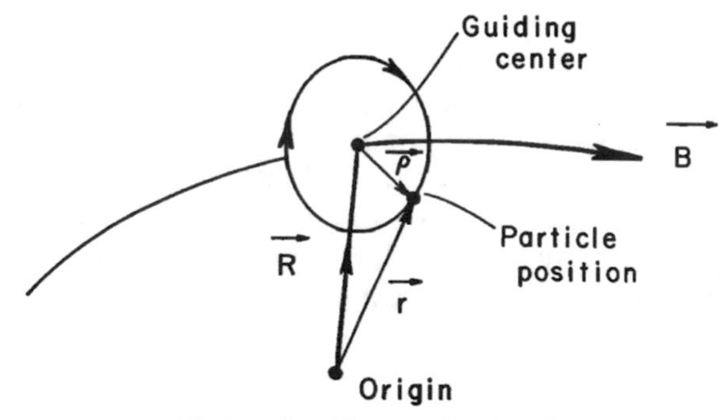

Fig. 1.2 A figure from Northrop (1963). This figure depicts the gyromotion about the local magnetic field of a positively charged particle. Image copyright: American Geophysical Union (reproduced with permission)

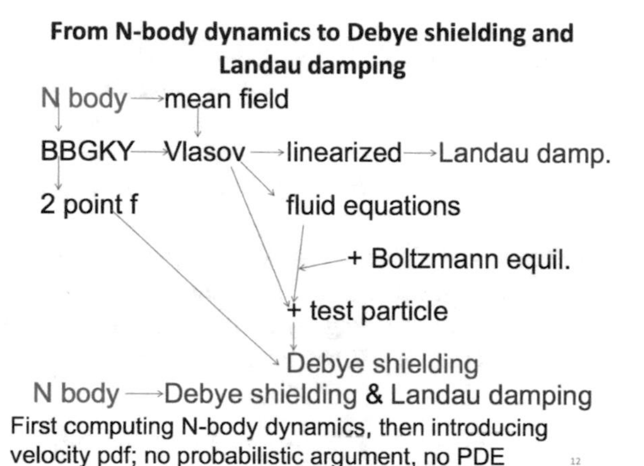

Fig. 1.3 A representation of the different models, approaches and phenomena in plasma physics. Image copyright: Dominique Escande: From his presentation at the EPS Conference on Plasma Physics 2015 in Lisbon (reproduced with permission)

for the gyroradius 'vector' and magnitude respectively (in contrast to the use of r_L herein); $\boldsymbol{r}$ is the particle position; and $\boldsymbol{R}$ is the guiding center position, such that $\boldsymbol{r} = \boldsymbol{R} + \boldsymbol{\rho}$. The local gyromotion is governed by the conservation (to lowest order) of the *magnetic moment*,

$$\mu = \frac{m_s v_\perp^2}{2|\boldsymbol{B}|},$$

for m_s the mass of a particle, and $v_\perp^2$ the square magnitude of the particle velocity normal to the local magnetic field. This theory is very useful for heuristic understanding of individual particle motion, and for the study of 'test particles' embedded in a system of interest (e.g. see Threlfall et al. 2015; Borissov et al. 2016), however not for 'building up' a theory that models the evolution of the particles and electromagnetic fields self-consistently. In a situation in which many particles are present, the self-consistent modelling of all of the particles would in practice require knowledge of the individual particle interactions via the electromagnetic fields of mixed origin (microscopic/self-generated and macroscopic/external fields), and in principle collisions, which is mathematically unwieldy. However, we note here that it is possible—whilst unconventional—to use N-body particle dynamics to study collective effects in plasma physics (e.g. see Pines and Bohm 1952; Escande et al. 2016), including the recent work of Dominique Escande and collaborators, who have taken an N-body approach to 're-deriving' physical phenomena, such as Debye shielding and Landau Damping (see Fig. 1.3 for a representation of how their work 'sidesteps' the more traditional routes).

1.1.2 Kinetic Theory

To move forward we require a mean-field/statistical formalism that allows for a self-consistent set of evolution equations, involving the quantities that both describe the particles and electromagnetic fields. The electromagnetic fields are governed by Maxwell's equations, and given in free space as

$$\nabla \cdot \boldsymbol{E} = \frac{\sigma}{\epsilon_0}, \tag{1.2}$$

$$\nabla \times \boldsymbol{E} = -\frac{\partial \boldsymbol{B}}{\partial t}, \tag{1.3}$$

$$\nabla \times \boldsymbol{B} = \mu_0 \boldsymbol{j} + \frac{1}{c^2}\frac{\partial \boldsymbol{E}}{\partial t}, \tag{1.4}$$

$$\nabla \cdot \boldsymbol{B} = 0, \tag{1.5}$$

for σ and $\boldsymbol{j}$ the charge and current densities respectively (e.g. see Griffiths 2013). The electric permittivity and magnetic permeability in vacuo are given by ϵ_0 and μ_0 respectively, and they are related by $c^2 = 1/(\mu_0\epsilon_0)$, for c the speed of light in free space. The electric and magnetic fields are defined as derivatives of the electrostatic scalar potential, ϕ, and the magnetic vector potential, $\boldsymbol{A}$, according to

$$\boldsymbol{E} = -\nabla\phi - \frac{\partial \boldsymbol{A}}{\partial t}, \tag{1.6}$$

$$\boldsymbol{B} = \nabla \times \boldsymbol{A}. \tag{1.7}$$

The potential functions are themselves 'sourced' by σ and $\boldsymbol{j}$, respectively,

$$\phi(\boldsymbol{x}, t) = \frac{1}{4\pi\epsilon_0}\int_{-\infty}^{\infty}\int_{-\infty}^{\infty}\int_{-\infty}^{\infty}\frac{\sigma(\boldsymbol{x}', t_r)}{|\boldsymbol{x}-\boldsymbol{x}'|}d^3x', \tag{1.8}$$

$$\boldsymbol{A}(\boldsymbol{x}, t) = \frac{\mu_0}{4\pi}\int_{-\infty}^{\infty}\int_{-\infty}^{\infty}\int_{-\infty}^{\infty}\frac{\boldsymbol{j}(\boldsymbol{x}', t_r)}{|\boldsymbol{x}-\boldsymbol{x}'|}d^3x', \tag{1.9}$$

for $t_r = t - |\boldsymbol{x}-\boldsymbol{x}'|/c$ the *retarded time* (Griffiths 2013). The charge and current densities can be calculated by taking *moments* of the *1-particle distribution functions*

(DF), $f_s(\boldsymbol{x}, \boldsymbol{v}; t)$ for particle species s (e.g. see Krall and Trivelpiece 1973; Schindler 2007), over velocity space

$$\sigma(\boldsymbol{x}, t) = \sum_s q_s n_s = \sum_s q_s \int_{-\infty}^{\infty}\int_{-\infty}^{\infty}\int_{-\infty}^{\infty} f_s d^3 v, \tag{1.10}$$

$$\boldsymbol{j}(\boldsymbol{x}, t) = \sum_s q_s n_s \boldsymbol{V}_s = \sum_s q_s \int_{-\infty}^{\infty}\int_{-\infty}^{\infty}\int_{-\infty}^{\infty} \boldsymbol{v} f_s d^3 v, \tag{1.11}$$

with $\boldsymbol{n}_s$ and $\boldsymbol{V}_s$ the number density and bulk velocity of particle species s respectively. Hereafter we us the notation d^3x and d^3v to imply triple integration over all position and velocity space respectively,

$$\int d^3x := \int_{-\infty}^{\infty}\int_{-\infty}^{\infty}\int_{-\infty}^{\infty} d^3x,$$
$$\int d^3v := \int_{-\infty}^{\infty}\int_{-\infty}^{\infty}\int_{-\infty}^{\infty} d^3v,$$

unless otherwise stated. The DF, f_s, represents the number density of particles in a microscopic volume of six-dimensional phase-space at a particular time, such that

$$f_s(\boldsymbol{x}, \boldsymbol{v}; t)d^3x d^3v = \#\text{ of particles in volume } d^3x \text{ centred on } \boldsymbol{x}$$
$$\text{with velocities in the range } (\boldsymbol{v},\ \boldsymbol{v} + d\boldsymbol{v}).$$

Note that one can instead use the Klimontovich-Dupree description to exactly describe the particles using Dirac-Delta functions in phase space, but this approach is really only useful for formal considerations (Krall and Trivelpiece 1973).

Now we are in a position to imagine the 'machine' behind nature's self-consistent evolution of the particles and fields in the plasma, in the following way:

Statistical description: $f_s(\boldsymbol{x}, \boldsymbol{v}, t_r)$ is found by 'coarse graining' (or 'ensemble averaging') the exact positions and velocities of the particles of species s at time t_r (Krall and Trivelpiece 1973; Fitzpatrick 2014)
Source terms: $\sigma(\boldsymbol{x}, t_r)$ and $\boldsymbol{j}(\boldsymbol{x}, t_r)$ are then found by integrating $f_s(\boldsymbol{x}, \boldsymbol{v}, t_r)$ over velocity space (Eqs. 1.10 and 1.11)
Potentials: $\phi(\boldsymbol{x}, t)$ and $\boldsymbol{A}(\boldsymbol{x}, t)$ are found by integrating $\sigma(\boldsymbol{x}, t_r)$ and $\boldsymbol{j}(\boldsymbol{x}, t_r)$ (Eqs. 1.8 and 1.9)
Forces: $\boldsymbol{F}_s(t)$ is found by differentiating the $\phi(\boldsymbol{x}, t)$ and $\boldsymbol{A}(\boldsymbol{x}, t)$ (Eqs. 1.6 and 1.7)
Velocities: $\boldsymbol{v}(t + \delta t)$ is found by integrating the Lorentz force, $\boldsymbol{F}_s(\boldsymbol{x}, t)$, for δt some infinitesimal time (Eq. 1.1)
Positions: $\boldsymbol{x}(t + 2\delta t)$ is found by integrating $\boldsymbol{v}(t + \delta t)$
Statistical description: $f_s(\boldsymbol{x}, \boldsymbol{v}, t + 2\delta t)$ is found by... and so the cycle continues.

To put these ideas on a firm mathematical footing, we need to understand the evolution of f_s in phase space, $(\boldsymbol{x}, \boldsymbol{v}; t)$. The DF evolves according to an equation typically known as the *Boltzmann equation*,

$$\frac{\partial f_s}{\partial t} + \boldsymbol{v} \cdot \frac{\partial f_s}{\partial \boldsymbol{x}} + \frac{q_s}{m_s} (\boldsymbol{E} + \boldsymbol{v} \times \boldsymbol{B}) \cdot \frac{\partial f_s}{\partial \boldsymbol{v}} = \left. \frac{\partial f_s}{\partial t} \right|_c , \tag{1.12}$$

with the right-hand side (RHS) of the equation describing the evolution of the DF according to 'collisions' (e.g. binary Coulomb collsions, see Fitzpatrick 2014). Properly, this equation is specifically named after the form of collision operator assumed, e.g. Boltzmann, Fokker-Planck or Lenard-Balescu (Schindler 2007). If the collision operator chosen is a function of f_s alone, then the Boltzmann equation and Maxwell's equations form a closed set, and the plasma is said to be in a *kinetic regime* Schindler (2007). In its general form, the Boltzmann equation can be obtained by integrating the Liouville equation for the N-particle DF in $6N$ dimensional phase-space,

$$\frac{dF_s(\boldsymbol{x}_1, \ldots, \boldsymbol{x}_N, \boldsymbol{v}_1, \ldots, \boldsymbol{v}_N; t)}{dt} = 0,$$

over the positions and velocities of all but one particle (Krall and Trivelpiece 1973) (made possible by the fact that particles of a particular species are identical (Tong 2012)). This also involves some assumptions made about the weak nature of the particle coupling in the plasma, characterised by

$$g = \frac{4\pi}{3\Lambda_p} = \frac{1}{n_e \lambda_D^3} \ll 1,$$

for the small parameter g, i.e. a *weakly coupled plasma* (Schindler 2007; Krall and Trivelpiece 1973). Here, Λ_p is the *plasma parameter*, equal to the number of electrons in the *Debye sphere*, a sphere of radius λ_D beyond which charge density inhomogeneities are shielded (Krall and Trivelpiece 1973; Fitzpatrick 2014). The small parameter g is used as the ordering parameter in an infinite hierarchy of statistical equations—the so called *BBGKY hierarchy*—for which closure is achieved by neglecting terms of the desired order in g^s (Krall and Trivelpiece 1973). The standard collisional framework is achieved by neglecting terms of order g^2 and above.

1.1.3 Quasineutrality

It is a feature common to many weakly coupled plasmas that typical spatial variations, L, are much larger than a quantity known as the *Debye radius*, λ_D,

$$\epsilon = \frac{\lambda_D}{L} \ll 1, \quad \text{s.t.} \quad \lambda_D = \sqrt{\frac{\epsilon_0 k_B T_e}{n e^2}},$$

for k_B Boltzmann's constant, T_e the electron temperature, and e the fundamental charge. In such a situation the plasma is considered to be *quasineutral* (Schindler 2007), typically taken to mean that

$$n_i = n_e \iff \sigma = 0. \tag{1.13}$$

Note that this is in an asymptotic sense, and formally does not imply that $\nabla \cdot \boldsymbol{E}$ vanishes, see e.g. Freidberg (1987), Schindler (2007), Harrison and Neukirch (2009b). To see how this works, first notice that if one normalises Poisson's equation by

$$\phi = \phi_0 \tilde{\phi}, \quad \nabla = \frac{1}{L}\tilde{\nabla}, \quad \sigma = e n_0 \tilde{\sigma},$$

for characteristic values ϕ_0, L and n_0 of the scalar potential, length scales and number densities, then one obtains

$$\epsilon^2 \tilde{\nabla}^2 \tilde{\phi} = -\tilde{\sigma},$$

for $\phi_0 = k_B T_0/e$, and $\epsilon = \lambda_D/L$. In the quasineutral limit the ϵ^2 parameter is vanishingly small. If one then makes an expansion of small parameters

$$\tilde{\phi} = \sum_{n=0}^{\infty} \epsilon^{2n} \tilde{\phi}_n, \quad \tilde{\sigma} = \sum_{n=0}^{\infty} \epsilon^{2n} \tilde{\sigma}_n,$$

then one sees that formally, for $\lambda_D/L \ll 1$,

$$\begin{aligned} \tilde{\sigma}_0 &= 0, \\ \tilde{\nabla}^2 \tilde{\phi}_0 &= -\tilde{\sigma}_1. \\ &\vdots \end{aligned}$$

As such, letting $\sigma = 0$ is an approximation to the quasineutral limit, valid to *first order*.

It should also be mentioned that quasineutrality implies that the characteristic frequencies are much less than the (electron) plasma frequency,

$$\omega_p = \sqrt{\frac{n_e e^2}{\epsilon_0 m_e}}. \tag{1.14}$$

Quoting Freidberg (1987) directly: "*For any low-frequency macroscopic charge separation that tends to develop, the electrons have more than an adequate time to respond, thus creating an electric field which maintains the plasma in local quasineutrality*". The assumption of quasineutrality is consistent with neglecting the displacement current in Maxwell's equations (Schindler 2007). These ordering assumptions give the quasineutral 'low-frequency/pre-Maxwell' equations that are commonly used in plasma physics

$$\nabla \times \boldsymbol{B} = \mu_0 \boldsymbol{j}, \qquad \text{Ampère's Law}$$

$$\nabla \times \boldsymbol{E} = -\frac{\partial \boldsymbol{B}}{\partial t}, \qquad \text{Faraday's Law}$$

$$\nabla \cdot \boldsymbol{B} = 0 \qquad \text{Solenoidal constraint,}$$

and

$$\left(\nabla \cdot \boldsymbol{E} = \frac{\sigma}{\epsilon_0}, \quad \text{Gauß' Law, s.t.} \quad \frac{\epsilon_0 \nabla \cdot \boldsymbol{E}}{\sigma} \ll 1 \right).$$

In practice, Gauß' Law is often not considered as a 'core equation' in plasma physics, and is implicitly 'replaced' by $\sigma = 0$. Faraday's law is also often 'reformulated' by eliminating the electric field using some version of Ohm's law (e.g. see Schindler 2007; Kulsrud 1983; Freidberg 1987; Krall and Trivelpiece 1973; Fitzpatrick 2014).

1.1.4 Fluid Models

Fluid models are the next step in the hierarchy after kinetic models, and are characterised by variables that depend only on space and time. Hence, the fluid equations are calculated by integrating over velocity space: taking velocity space *moments* of the kinetic equation at hand (Schindler 2007). This process was laid down in the seminal work of Braginskii (1965), giving the collisional transport (or Braginskii) equations

$$\frac{\partial \rho_e}{\partial t} + \rho_e \nabla \cdot \boldsymbol{V}_e = 0, \qquad \textit{Electron mass transport}$$

$$\rho_e \frac{d\boldsymbol{V}_e}{dt} + \nabla p_e + \nabla \cdot \boldsymbol{\pi}_e - \sigma_e(\boldsymbol{E} + \boldsymbol{V}_e \times \boldsymbol{B}) = \boldsymbol{F}_{\mathrm{fr},e}, \qquad \textit{Electron mom. transport}$$

$$\frac{3}{2}\frac{dp_e}{dt} + \frac{5}{2} p_e \nabla \cdot \boldsymbol{V}_e + \boldsymbol{\pi}_e : \nabla \boldsymbol{V}_e + \nabla \cdot \boldsymbol{q}_e = W_e, \qquad \textit{Electron energy transport}$$

for electrons, and

$$\frac{\partial \rho_i}{\partial t} + \rho_i \nabla \cdot \boldsymbol{V}_i = 0, \qquad \textit{Ion mass transport}$$

$$\rho_i \frac{d\boldsymbol{V}_i}{dt} + \nabla p_i + \nabla \cdot \boldsymbol{\pi}_i - \sigma_i(\boldsymbol{E} + \boldsymbol{V}_i \times \boldsymbol{B}) = -\boldsymbol{F}_{\mathrm{fr},i}, \qquad \textit{Ion mom. transport}$$

$$\frac{3}{2}\frac{dp_i}{dt} + \frac{5}{2} p_i \nabla \cdot \boldsymbol{V}_i + \boldsymbol{\pi}_i : \nabla \boldsymbol{V}_i + \nabla \cdot \boldsymbol{q}_i = W_e, \qquad \textit{Ion energy transport}$$

for ions, using the notation from Fitzpatrick (2014). In these equations $\rho_s = m_s n_s$ defines the mass density, $p_s = \frac{1}{3}\mathrm{Tr}(\boldsymbol{P}_s)$ the scalar pressure for species s, defined by the trace of the pressure tensor of species s

$$P_{ij,s} = m_s \sum_s \int f_s w_{is} w_{js} d^3 v \quad \text{s.t.} \quad P_{ij} = \sum_s P_{ij,s},$$

for $\boldsymbol{w}_s = \boldsymbol{v} - \boldsymbol{V}_s$ the velocity of a particle relative to the bulk flow, and for which

$$\boldsymbol{\pi}_s = \boldsymbol{P}_s - p_s \boldsymbol{I},$$

is the stress/generalised viscosity tensor. The vector $\boldsymbol{q}_s$,

$$\boldsymbol{q}_s = \frac{m_s}{2} \int w_s^2 \boldsymbol{w}_s f_s d^3 v,$$

is the heat flux density. Finally, $\boldsymbol{F}_{\mathrm{fr},s}$ and W_s are found by taking the momentum- and energy- moments of the collision operator (the RHS of the Boltzmann equation), and represent the collisional friction force, and collisional energy change, respectively.

These are the *two-fluid* equations. They describe the spatio-temporal evolution of the moments of the ion and electron DFs resepctively, and these are coupled by the EM fields. In their current form they are not closed: there are more unknowns than equations (Freidberg 1987). It is not the purpose of this introduction to explore the subtle details of fluid closure, two-fluid, single fluid and magnetohydrodynamic (MHD) theories. For details on these topics see Schindler (2007), Kulsrud (1983), Freidberg (1987), Krall and Trivelpiece (1973), and Fitzpatrick (2014).

1.2 Collisions in Plasmas

1.2.1 *Collisional Plasmas*

The collisionality of a plasma species is characterised in time and space by two quantities (Fitzpatrick 2014): the collision rate/frequency, ν_s; and the mean free path $\lambda_{\mathrm{mfp},s}$, such that

$$\begin{aligned} \nu_s &\approx \sum_{s'} \nu_{ss'}, \\ \lambda_{\mathrm{mfp},s} &= v_{\mathrm{th},s}/\nu_s, \\ T_i = T_e &\implies \nu_e \sim \sqrt{\frac{m_i}{m_e}}\nu_i. \end{aligned}$$

That is to say that the total collision rate for a species is made up of the collision rates with all species (including its own), the mean free path measures the typical distance a particle travels between collisions, and that in the case of an isothermal plasma the collision rate for electrons is much greater than that for ions. The thermal velocity, $v_{\text{th},s}$, gives the energy of random particle motion $E_{\text{random}} = m_s v_{\text{th},s}^2$, such that in thermal equilibrium $k_B T_s = E_{\text{random}}$ (Schindler 2007). We note here that a collision is classified as a $\geq 90°$ scattering event, and as such a particle may have numerous 'small-angle' scattering (i.e. $<90°$) events before a successful 'collision' (Fitzpatrick 2014).

A collision dominated plasma is one for which the mean free path is much smaller than typical plasma length scales, L

$$\lambda_{\text{mfp}} \ll L,$$

with the opposite limit indicating a collisionless plasma. The collisional frequency typically has magnitude

$$\nu_e \sim \frac{\ln \Lambda_p}{\Lambda_p} \omega_p,$$

(Fitzpatrick 2014) and as such

$$\nu_e \ll \omega_p \iff \Lambda_p \gg 1 \iff g \ll 1.$$

That is to say that weakly coupled plasmas are those for which collisions are not able to prevent plasma oscillations from regulating charge separation. In the case of a sufficiently collisional plasma characterised by

$$\begin{aligned} \frac{1}{\nu_s}\frac{\partial \langle v^k f_s \rangle}{\partial t} &\ll \langle v^k f_s \rangle, \\ \lambda_{\text{mfp},s} \nabla \langle v^k f_s \rangle &\ll \langle v^k f_s \rangle, \\ \lambda_{\text{mfp},s} e |\boldsymbol{E}| &\ll k_B T_s \end{aligned}$$

for which $\langle v^k f_s \rangle$ is a k-th order velocity moment of the DF, then the plasma is in a *local thermal equilibrium* (e.g. see Cowley 2003/4), characterised by a temperature $T_s(\boldsymbol{x}, t)$, and the DF can be written as a Maxwellian of the form

$$f_s(\boldsymbol{x}, \boldsymbol{v}; t) = \frac{n_s(\boldsymbol{x}; t)}{(2\pi k_B T_s(\boldsymbol{x}; t)/m_s)^{3/2}} e^{-m_s(\boldsymbol{v}-\boldsymbol{V}_s(\boldsymbol{x};t))^2/(k_B T_s(\boldsymbol{x};t))}, \tag{1.15}$$

to lowest order. This DF describes a plasma species with local number density $n_s(\boldsymbol{x}, t)$ and local bulk velocity $\boldsymbol{V}_s(\boldsymbol{x}, t)$. The DF in Eq. (1.15) is clearly not an equilibrium solution, since the number density, bulk flow and temperature explicitly depend on time. Given sufficient time, Boltzmann's *H-Theorem* implies that collisions will always attempt to drive a system towards thermal equilibrium (e.g. see Grad 1949; Brush 2003), defined by a DF of the form

$$f_s(\boldsymbol{v}) = \frac{n_s}{(2\pi k_B T_s/m_s)^{3/2}} e^{-m_s(\boldsymbol{v}-\boldsymbol{V}_s)^2/(k_B T_s)}. \tag{1.16}$$

The DF in Eq. (1.16) is of the same form as that in Eq. (1.15), but is now independent of space and time. The temperature is constant and a non-zero bulk flow is permitted.

1.2.2 Collisionless Plasmas

The statement that collisionless plasmas are those for which $\lambda_{\mathrm{mfp}} \gg L$ is rather truistic, and not particularly helpful in physical terms. Using the definition of the plasma parameter (Fitzpatrick 2014),

$$\Lambda_p = \frac{4\pi}{n_e^{1/2}} \left(\frac{\sqrt{\epsilon_0 T_e}}{e} \right)^3,$$

we see that the collision frequency behaves like

$$\nu_e \sim \frac{e^4 n_e \ln \Lambda_p}{4\pi\epsilon_0^2 m^{1/2} T_e^{3/2}} = \frac{e^4}{4\pi\epsilon_0^2 m^{1/2}} \frac{n_e}{T_e^{3/2}} \ln \left(\frac{4\pi}{n_e^{1/2}} \left(\frac{\sqrt{\epsilon_0 T_e}}{e} \right)^3 \right).$$

Hence, dense and low temperature plasmas are more likely to be collisional, whereas diffuse and high temperature plasmas tend to be collisionless. In such situations, it is reasonable to neglect the RHS of the Boltzmann equation (Eq. (1.12)), giving the *Vlasov equation* (Vlasov 1968),

$$\frac{\partial f_s}{\partial t} + \boldsymbol{v} \cdot \frac{\partial f_s}{\partial \boldsymbol{x}} + \frac{q_s}{m_s} (\boldsymbol{E} + \boldsymbol{v} \times \boldsymbol{B}) \cdot \frac{\partial f_s}{\partial \boldsymbol{v}} = 0. \tag{1.17}$$

In closed form this equation can be written, using Hamilton's equations (Tong 2012), as

$$\begin{aligned} \frac{df_s}{dt} &= \frac{\partial f_s}{\partial t} + \frac{\partial f_s}{\partial \boldsymbol{x}} \cdot \frac{d\boldsymbol{x}}{dt} + \frac{\partial f_s}{\partial \boldsymbol{v}} \cdot \frac{d\boldsymbol{v}}{dt} = 0, \\ &= \frac{\partial f_s}{\partial t} + \frac{\partial f_s}{\partial \boldsymbol{x}} \cdot \frac{\partial H_s}{\partial \boldsymbol{p}_s} - \frac{\partial f_s}{\partial \boldsymbol{p}_s} \cdot \frac{\partial H_s}{\partial \boldsymbol{x}} = 0, \\ &= \frac{\partial f_s}{\partial t} + \{f_s, H_s\}_{PB} = 0. \end{aligned} \tag{1.18}$$

Here, the Hamiltonian is given by H_s, the canonical momenta by $\boldsymbol{p}_s$, and the brackets $\{\,,\,\}_{PB}$ are Poisson brackets, whose definition can be inferred from above. We can go from using velocity variables in the first line, to momentum variables in the second since $d\boldsymbol{p}_s = m_s d\boldsymbol{v}$. The Vlasov equation essentially states that the DF is conserved along a particle trajectory in phase-space (Schindler 2007), since the characteristics of the Vlasov equation are the single particle equations of motion,

$$\frac{d}{dt}\boldsymbol{x}(t) = \boldsymbol{v}(t),$$
$$\frac{d}{dt}\boldsymbol{v}(t) = \frac{q_s}{m_s}(\boldsymbol{E} + \boldsymbol{v} \times \boldsymbol{B}).$$

The solutions of this equation are in principle completely reversible in time, and hence entropy conserving (Krall and Trivelpiece 1973).

1.3 Collisionless Plasma Equilibria

A Vlasov equilibrium is obtained when the DF satisfies

$$\frac{\partial f_s}{\partial t} = 0 \implies \{H_s, f_s\}_{PB} = 0. \tag{1.19}$$

This statement does not mean that there are no macroscopic particle flows or currents; density, pressure or temperature gradients; or even heat fluxes, for example. That is to say that the moments of the DF can still have gradients in space. Rather, it is an equilibrium in the sense of a particle distribution. This means that the value of the DF at each individual point in phase-space is independent of time.

It is a standard result in classical mechanics that constants of motion, $C_s(\boldsymbol{x}(t), \boldsymbol{p}(t))$, (that do not depend explicitly on time) are in 'involution' with/commute with the Hamiltonian (Tong 2004),

$$\{H_s, C_s\}_{PB} = 0. \tag{1.20}$$

Using this result, and the linearity of the Poisson bracket, we see that any function of the constants of motion is a Vlasov equilibrium DF, since

$$\{H_s, f_s(C_{1s}, \ldots, C_{ns})\}_{PB} = \sum_{j=1}^{n} \frac{\partial f_s}{\partial C_{js}} \left\{H_s, C_{js}\right\}_{PB} = \sum_{j=1}^{n} \frac{\partial f_s}{\partial C_{js}} \times 0 = 0. \tag{1.21}$$

We can also show that the reverse is true, namely that any Vlasov equilibrium DF is a function of the constants of motion. First consider a Vlasov equilibrium DF $f_s(G_1, G_2, \ldots, G_n)$ for arbitrary linearly independent functions $G_j(\boldsymbol{x}(t), \boldsymbol{p}(t))$. Then by linearity of the Poisson Bracket,

$$\{H_s, f_s\}_{PB} = \sum_{j=1}^{n} \frac{\partial f_s}{\partial G_j} \left\{H_s, G_j\right\}_{PB}. \tag{1.22}$$

This sum must be zero for an equilibrium, and since the G_j are linearly independent, that implies that each of the Poisson brackets must be zero independently. Hence the G_j must be constants of motion and so

$$\text{"}f_s \textit{ is a Vlasov equilibrium DF} \iff f_s \textit{ is a function of the constants of motion".}$$

It is clear that a Vlasov equilibrium DF also satisfies the time-dependent Vlasov equation itself Schindler (2007), since

$$\frac{df_s}{dt} = \frac{\partial f_s}{\partial t} + \{f_s, H_s\}_{PB} = 0 + 0. \tag{1.23}$$

Using this fact, one can construct time-dependent solutions for 'nonlinear' propagating structures to the Vlasov equation by using a frame transformation (Schamel 1979). Then one can solve for Vlasov equilibria in the wave frame, e.g. the famous BGK modes (Bernstein et al. 1957) and Schamel's theory (Schamel 1986), amongst other examples, e.g. see Abraham-Shrauner (1968), Ng and Bhattacharjee (2005), Vasko et al. (2016), Hutchinson (2017).

1.3.1 The 'Forward' and 'Inverse' Approaches

As described above, one can easily construct equilibrium solutions of the Vlasov equation provided that at least one constant of motion has been identified. Any differentiable function of the constants of motion is an equilibrium solution of the Vlasov equation (Schindler 2007), and is physically meaningful provided all velocity moments exist,

$$\left| \int v_1^i v_2^j v_3^k \, f_s \, dv_1 \, dv_2 \, dv_3 \right| < \infty \, \forall i, \; j, \; k \in 0, 1, 2, \dots ,$$

and the function is non-negative over all phase-space,

$$f_s(\boldsymbol{x}, \boldsymbol{v}) \geq 0 \, \forall \boldsymbol{x}, \; \boldsymbol{v}.$$

Whilst such a function may well satisfy these mathematical/microscopic conditions, the next question to ask is of the macroscopic electromagnetic fields that are consistent with such a function. Through Eqs. (1.10) and (1.11), we see that the distribution of particles in phase-space determines the charge and current densities respectively, in configuration-space. These charge and current densities are consistent with certain electric and magnetic fields through Maxwell's equations (Eqs. (1.2)–(1.3)). Hence, a full understanding of the macroscopic and microscopic physics of a plasma necessitates a self-consistent 'solution' of the Vlasov-Maxwell (VM) system.

From these considerations, it should be clear that there are two possible routes to follow, in the absence of a comprehensive self-consistent theory, namely

- 'Inverse': Given some or all of the macroscopic fields (ϕ, $\boldsymbol{A}$), can we find a self-consistent DF, f_s? (e.g. see discussions in Alpers 1969; Channell 1976; Mynick et al. 1979; Greene 1993; Harrison and Neukirch 2009b; Belmont et al. 2012; Allanson et al. 2016)
- 'Forward': Given a DF, f_s, can we find some set of self-consistent macroscopic fields, (ϕ, $\boldsymbol{A}$)? (e.g. see discussions in Grad 1961; Harris 1962; Sestero 1964, 1965; Lee and Kan 1979a; Schindler 2007; Kocharovsky et al. 2010; Vasko et al. 2013)

The forward approach is the one that is most frequently seen in the literature. This is partly due, mathematically, to the fact that this involves solving differential equations, as opposed to the often less tractable inversion of integral equations in the case of the inverse approach. But also, as argued in Sect. 1.2.1, it is reasonable on physical grounds to assume that—for sufficiently collisional (Cowley 2003/4) and 'not-too-turbulent' plasmas (Alpers 1969)—that the DF is (locally) Maxwellian, and then to proceed with the forwards approach from thereon.

In the case of collisionless plasmas, there are an infinite class of equilibrium solutions in principle, and hence the forwards approach would have to be predicated on some prior knowledge of the DF. In-situ observations of DFs have only recently become available with spatio-temporal resolution on kinetic scales, for example the NASA Multiscale Magnetospheric (MMS) mission (Hesse et al. 2016), and the ESA candidate mission: Turbulent Heating ObserveR (THOR) (Vaivads et al. 2016).

Due to the ubiquitous nature and reasonable validity of the MHD approach in many environments, and the relative wealth and long history of magnetic field measurements, the equilibrium structures and dynamics of electromagnetic fields are better understood and more often used as the fundamental basis, or object, of plasma physics discussions and theory. Hence, it is of use, and necessity, to consider the inverse approach.

1.3.2 Motivating Translationally Invariant Vlasov-Maxwell (VM) Equilibria

1.3.2.1 Current Sheets

In a planar geometry, localised electric currents in a plasma are known as current sheets: frequently considered to be the initial state of wave processes (Fruit et al. 2002), instabilities (Schindler 2007), reconnection (Yamada et al. 2010) and various dynamical phenomena in laboratory (Beidler and Cassak 2011), space (Zelenyi et al. 2011) and astrophysical (DeVore et al. 2015) plasmas. The formation of current sheets is ubiquitous in plasmas. They can form between plasmas of different origins that encounter each other, such as at Earth's magnetopause between the magnetosheath

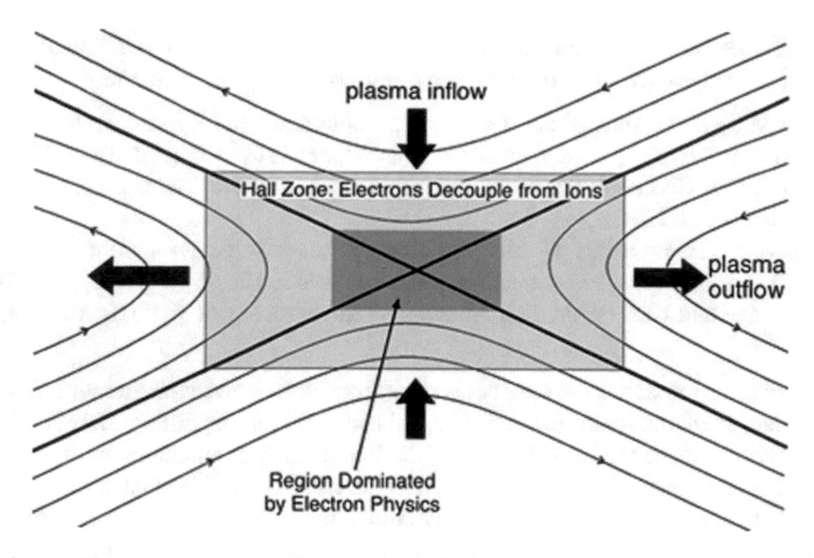

Fig. 1.4 A diagrammatic representation of the local structure of a magnetic reconnection event, and the 'electron diffusion region', in which the electrons decouple from the magnetic field. Image copyright: NASA MMS-SMART Investigation, (reproduced with permission)

plasmas and magnetospheric plasmas (e.g. see Dungey 1961; Phan and Paschmann 1996); or they can develop spontaneously in magnetic fields that are subjected to random external driving (e.g. see Parker 1994), such as in the solar corona.

As to be introduced in Sect. 1.3.3, localised electric currents are an important ingredient for magnetic reconnection: acting as a signature of sheared magnetic fields, and reconnection electric fields (e.g. see Biskamp 2000; Hesse et al. 2011). As per Poynting's theorem (Poynting 1884), with $\boldsymbol{S} = \mu_0^{-1}\boldsymbol{E} \times \boldsymbol{B}$, and neglecting electric field energy,

$$\frac{\partial B^2/((2\mu_0)}{\partial t} = -\nabla \cdot \boldsymbol{S} - \boldsymbol{j} \cdot \boldsymbol{E},$$

intense current sheets are ideal locations for magnetic energy conversion and dissipation (Birn and Hesse 2010; Zenitani et al. 2011). The dominant mechanisms that release the free energy include magnetic reconnection, and various plasma instabilities.

The currents themselves are usually considered synonymous with a stressed and/or anti-parallel magnetic field configuration, since in a quasineutral plasma (or a plasma in equilibrium), the current density is given by

$$\boldsymbol{j} = \frac{1}{\mu_0}\nabla \times \boldsymbol{B}.$$

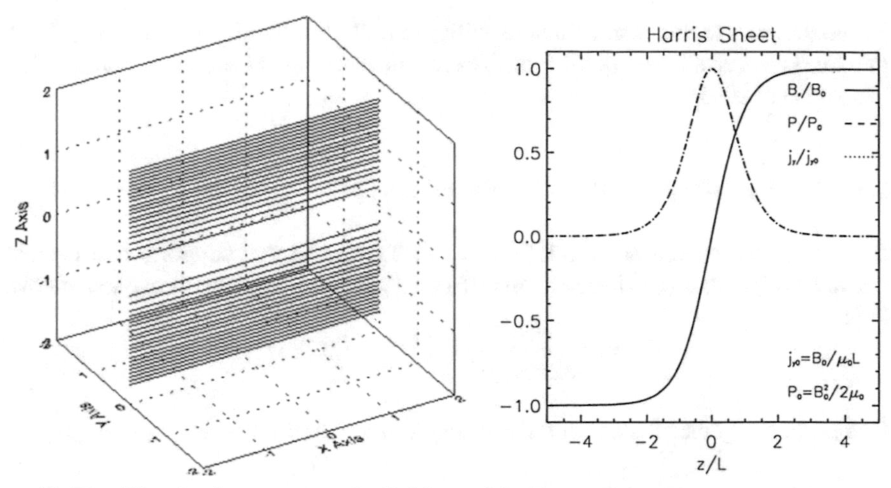

(a) The Harris sheet magnetic field (b) The Harris sheet equilibrium magnetic field, current density and scalar pressure.

Fig. 1.5 **a** represents the magnetic field lines for the Harris sheet magnetic field. **b** shows the normalised B_x, j_y, and scalar pressure p for the Harris sheet equilibrium characterised by $j_y = dB_x/dz$, and $dp/dz = -j_y B_x$. Image's copyright: M. G. Harrison's Ph.D thesis (Harrison 2009), (reproduced with permission)

Perhaps the most used current sheet equilibrium model is represented in Fig. 1.5: the Harris sheet (Harris 1962),

$$\begin{aligned} \boldsymbol{B} &= B_0\left(\tanh\left(\frac{z}{L}\right), 0, 0\right), \\ \frac{1}{\mu_0}\nabla\times\boldsymbol{B} = \boldsymbol{j} &= \frac{B_0}{\mu_0 L}\left(0, \operatorname{sech}^2\left(\frac{z}{L}\right), 0\right), \\ \frac{dp}{dz} = -j_y B_x \implies p &= \frac{B_0^2}{2\mu_0}\operatorname{sech}^2(z/L), \end{aligned} \tag{1.24}$$

with L the current sheet 'width', normalising z; B_0 the asymptotic values of the magnetic field, normalising B_x; $j_{y0} = B_0/(\mu_0 L)$ and $p_0 = B_0^2/(2\mu_0)$ normalising the current density and scalar pressure respectively. The maximum shear of B_x is localised in the region $-L < z < L$, and this is where we see the maximum values of the current density: the current sheet itself. A Vlasov equilibrium DF self-consistent with the Harris sheet is given by

$$f_s = \frac{n_{0s}}{(\sqrt{2\pi}v_{\mathrm{th},s})^3}e^{-\beta_s(H_s - u_{ys}p_{ys})}, \tag{1.25}$$

with $\beta_s = 1/(m_s v_{\mathrm{th},s}^2)$; n_{0s} a constant with dimensions of spatial number density (and not necessarily representing the number density itself); and with u_{ys} a bulk flow parameter, that in this case coincides with the bulk flow itself, i.e. $u_{ys} = V_{ys}$. Note that one can derive other equilibrium DFs for the Harris sheet, e.g. the Kappa (κ) DF (Fu and Hau 2005).

1.3.2.2 Harris-Type Distribution Functions (DFs)

If we were to 'generalise' the DF in Eq. (1.25) to one that supports two current density components (and hence a DF self-consistent with a different magnetic field), then we have

$$f_s = \frac{n_{0s}}{(\sqrt{2\pi}v_{\mathrm{th},s})^3}e^{-\beta_s(H_s - u_{xs}p_{xs} - u_{ys}p_{ys})}.$$

One particularly nice feature of a DF that is a function of $(H_s - u_{xs}p_{xs} - u_{ys}p_{ys})$,

$$f_s = f_s(H_s - u_{xs}p_{xs} - u_{ys}p_{ys})$$

is that the bulk flows are directly related to the flow parameters, i.e. $V_{xs} = u_{xs}$ and $V_{ys} = u_{ys}$. This is seen by the following argument. If we define $\mathcal{H}_s = H_s - \boldsymbol{u}_s \cdot \boldsymbol{p}_s$ for

$$\boldsymbol{u}_s = (u_{xs}, u_{ys}, 0), \quad \boldsymbol{p}_s = (p_{xs}, p_{ys}, 0),$$

then $f_s = f_s(\mathcal{H}_s)$ and

$$\mathcal{H}_s = \frac{m_s}{2}\boldsymbol{\mathcal{U}}_s^2 - \frac{m_s}{2}\boldsymbol{u}_s^2 - q_s(A_x + A_y) \quad \text{s.t.} \quad \boldsymbol{\mathcal{U}}_s = \boldsymbol{v} - \boldsymbol{u}_s.$$

If we now consider the first-order moment of f_s by $\boldsymbol{\mathcal{U}}_s$, the result must be zero since f_s only depends on $\boldsymbol{\mathcal{U}}_s^2$, through $\mathcal{H}_s$. Consequently

$$\int \boldsymbol{\mathcal{U}}_s f_s(\mathcal{H}_s) d^3\mathcal{U}_s = 0 = \underbrace{\int \boldsymbol{v} f_s d^3v}_{n_s\boldsymbol{V}_s} - \underbrace{\boldsymbol{u}_s \int f_s d^3v}_{n_s\boldsymbol{u}_s},$$

and hence $\boldsymbol{V}_s = \boldsymbol{u}_s = (u_{xs}, u_{ys}, 0)$.

1.3.2.3 Other Applications

Current sheets are by no means the only application of the work on translationally invariant VM equilibria in this thesis. As indicated in Sect. 1.3.6, translationally invariant VM equilibria are of use for numerous other applications in plasma physics. Examples include nonlinear waves (e.g. see Bernstein et al. 1957; Ng et al. 2012); electron holes, ion holes and double layers (e.g. see Schamel 1986); and colllisionless shock fronts (e.g. see Montgomery and Joyce 1969; Burgess and Scholer 2015).

1.3.3 Magnetic Reconnection

Magnetic reconnection is a ubiquitous phenomenon in solar, space, astrophysical and laboratory plasmas, and now considered to be "*among the most fundamental unifying concepts in astrophysics, comparable in scope and importance to the role of natural selection in biology.*" (Moore et al. 2015): see authoritative discussions of 'classical' reconnection in Schindler (2007, Priest and Forbes (2000), Biskamp (2000), Hesse et al. (2011); on modern theories of 'fast' reconnection and 'turbulent/stochastic reconnection' in Lazarian et al. (2015); Loureiro and Uzdensky (2016); and 'fractal reconnection' in Shibata and Tanuma (2001). The literature on the topic is vast and there are many complex concepts to consider regarding the precise mathematical definition (e.g. see Hesse and Schindler 1988; Priest 2014) of reconnection and its physical behaviour in different dimensions and plasma environments. The phenomenon also appears in physical environments as numerous as the number of plasma environments themselves, e.g. solar corona, planetary and pulsar magnetospheres, magnetic dynamos, gamma-ray bursts, geomagnetic storms and sawtooth crashes in tokamaks. However, there are common features that are agreed upon:

Topology: There is a change in the topology of the magnetic field, caused by processes in non-ideal ($\boldsymbol{E} + \boldsymbol{V} \times \boldsymbol{B} \neq \boldsymbol{0}$) regions of plasma with strong localised electric currents and parallel electric fields.

Diffusion region: This region is termed the diffusion region (e.g. see Hesse et al. 2001; Schindler 2007; Hesse et al. 2011), and is represented locally, and in an idealised geometry in Fig. 1.4.

Decoupling: Ideal MHD breaks down within the diffusion region, kinetic physics is dominant, and the plasma decouples from the magnetic field, enabling stored magnetic energy to be released to the physical medium.

Hence, magnetic reconnection explicitly couples (via the transmission of energy) the macroscopic ideal MHD picture of relatively slow-evolving and large scale neutral, conducting fluids to the small-scale, short-timescale and non-neutral kinetic plasma physics. Reconnection can of course occur in many different ways. It could occur in one of following ways

Incidental: One physical phenomenon out of many (and not necessarily dominant), occurring in a dynamical plasma, e.g. small scale reconnection in a turbulent plasma (e.g. Lazarian and Vishniac 1999);

Steady-state: A continuous reconnection phenomenon that generates kinetic energy with no significant macroscopic structural changes, e.g. the Sweet-Parker (Parker 1957; Sweet 1958) and Petschek models (Petschek 1964);

Instability: The result of an instability, i.e. the system was perturbed from equilibrium, reconnection was initiated, and the system does not return to the initial equilibrium, e.g. the tearing mode instability (e.g. see Furth et al. 1963; Drake and Lee 1977).

1.3.3.1 Approximate Equilibria In Particle-in-Cell (PIC) Simulations

Magnetic reconnection processes can critically depend on a variety of length and time scales, for example on lengths of the order of the Larmor orbits and below that of the mean free path (e.g. see Biskamp 2000; Birn and Priest 2007). In such situations a collisionless kinetic theory could be necessary to capture all of the relevant physics, and as such an understanding of the differences between using MHD, two-fluid, hybrid, Vlasov and other approaches is of paramount importance, for example see Birn et al. (2001, 2005) for discussions of this problem in the context of one-dimensional (1D) current sheets: the 'Geospace Environmnetal Modelling (GEM)' and 'Newton' challenges.

In the absence of an exact collisionless kinetic equilibrium solution, one has to use non-equilibrium DFs to start kinetic simulations, without knowing how far from the true equilibrium DF they are. In such cases, non-equilibrium drifting Maxwellian distributions are frequently used (see Swisdak et al. 2003; Hesse et al. 2005; Pritchett 2008; Malakit et al. 2010; Aunai et al. 2013; Hesse et al. 2013; Guo et al. 2014; Hesse et al. 2014; Liu and Hesse 2016 for examples),

$$f_{\mathrm{Maxw},s} = \frac{n_s(\boldsymbol{x})}{(\sqrt{2\pi}v_{\mathrm{th},s})^3}\exp\left[\frac{-(\boldsymbol{v}-\boldsymbol{V}_s(\boldsymbol{x}))^2}{2v_{\mathrm{th},s}^2}\right], \tag{1.26}$$

with $v_{\mathrm{th},s}$ a characteristic value of the thermal velocity, $n_s(\boldsymbol{x})$ the number density, and $\boldsymbol{V}_s$ the bulk velocity of species s. These DFs can reproduce the same moments n_s, $\boldsymbol{V}_s$ (and $p = n_s k_B T_s$, typically with $n_i = n_e$) necessary for a fluid equilibrium, maintained by the gradient of a scalar pressure,

$$\nabla p = \boldsymbol{j} \times \boldsymbol{B}.$$

However, the DF, $f_{\mathrm{Maxw,s}}$, in Eq. (1.26) is not an exact solution of the Vlasov equation and hence does not describe a kinetic equilibrium. The macroscopic force balance self-consistent with a quasineutral Vlasov/kinetic equilibrium is maintained by the divergence of a rank-2 pressure tensor, $P_{ij} = P_{ij}(A_x(z), A_y(z))$ (e.g. see Channell 1976; Mynick et al. 1979; Schindler 2007), according to

$$\nabla \cdot \boldsymbol{P} = \boldsymbol{j} \times \boldsymbol{B}.$$

As explained in Aunai et al. (2013) on the subject of PIC simulations, the fluid equilibrium characterised by a drifting Maxwellian can evolve to a quasi-steady state "*with an internal structure very different from the prescribed one*", and as demonstrated in Pritchett (2008), undesired electric fields, "*coherent bulk oscillations*", and other perturbations may form, in nature's attempt to maintain force-balance. Figure 1.6 is taken from Pritchett (2008), and demonstrates this phenomenon. Each of the panels relates, in principle, to a 1D MHD equilibrium characterised by $dp/dx = j_y B_z$, in which the PIC simulation is intialised with a DF of the form of that in Eq. (1.26). Panel (a) demonstrates how the initial condition is self-consistent with a magnetic

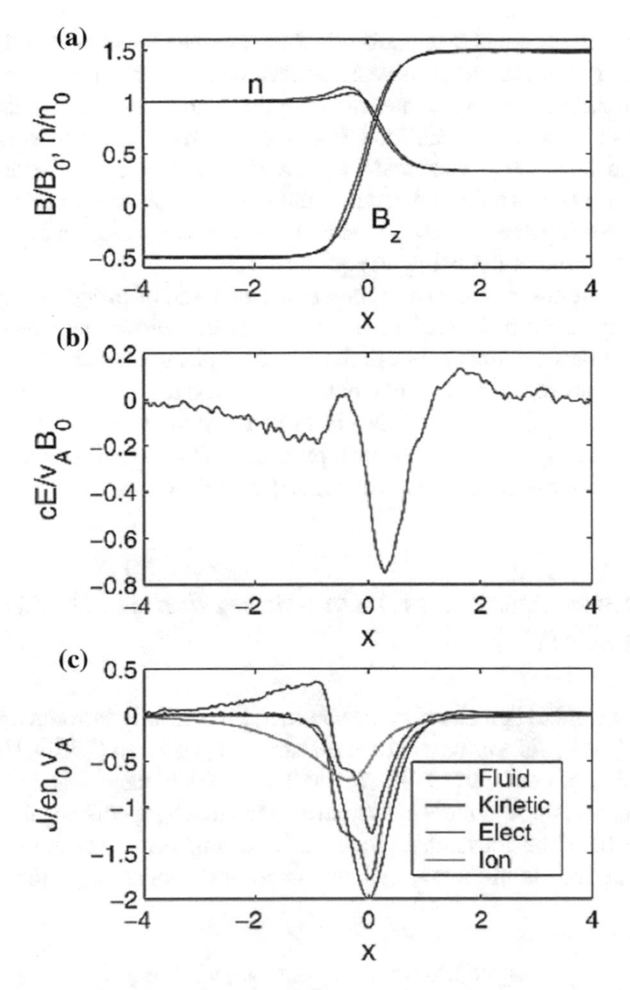

Fig. 1.6 A figure from Pritchett (2008). Profiles in x across a 1D current layer: **a** magnetic field $B_z(x)$ and density $n(x)$ from a PIC simulation at 'time' 20 (red curves), and from the fluid equilibrium (black curves); **b** electric field $E_x(x)$ from a PIC simulation at 'time' 20; **c** current density $J_y(x)$ determined from a PIC simulation at 'time' 20 carried by the electrons (blue curve), ions (green curve), and the electrons and ions combined (red curve) and the fluid current density corresponding to the magnetic field (black curve). Image copyright: American Geophysical Union (reproduced with permission)

field profile and number density that are very close to those prescribed by the fluid equilibrium. However, panel (b) shows an electric field that forms due to the non-equilibrium initial state, and panel (c) demonstrates the resultant disparity between the exact/'fluid' current density (black), and that derived from the PIC simulation (red).

The knowledge of exact VM equilibria thus provides the chance to initialise PIC simulations in full confidence, with the intended macroscopic quantities reproduced. Exact VM equilibria would also permit analytical and numerical studies of the linear phase of collisionless instabilities (Gary 2005), such as the tearing mode (e.g. see Drake and Lee 1977; Quest and Coroniti 1981a). This sort of exact analysis is formally out of reach without an exact initial condition since—as discussed by e.g. Pritchett (2008); Aunai et al. (2013)—a non-exact Vlasov solution creates perturbations itself, by virtue of not being an equilibrium.

Of course, one could make an argument on the basis of ordering arguments that a non-exact equilibrium DF such as that in Eq. (1.26) allows the study of the non-linear (and perhaps the linear) phase dynamics of plasma instabilities, such as the tearing mode. This sort of argument would be based on the assumption that a drifting Maxwellian such as that in Eq. (1.26) is *sufficiently* close to a VM equilibrium so as not to significantly affect the physical processes. However, it is generally unclear how far such an initial condition is from exact equilibrium.

1.3.4 Forward Approach for One-Dimensional (1D) VM Equilibria

To give context and to demonstrate the contrast, I will briefly introduce the 'forward approach' in VM equilibria, as used and discussed in e.g. Grad (1961), Harris (1962), Sestero (1967), Lee and Kan (1979a), Schindler (2007). In these—and other—works, a self-consistent solution to the VM system is found first by specifying the equilibrium DF as a function of the constants of motion. For example, a 1D system with $\partial/\partial x = \partial/\partial y = 0$, has the Hamiltonian, and two canonical momenta as the constants of motion,

$$H_s(\phi(\boldsymbol{x}), \boldsymbol{v}) = H_s(z, \boldsymbol{v}) = m_s \boldsymbol{v}^2/2 + q_s\phi(z), \tag{1.27}$$

$$p_{xs}(A_x(\boldsymbol{x}), \boldsymbol{v}) = p_{xs}(z, v_x) = m_s v_x + q_s A_x(z), \tag{1.28}$$

$$p_{ys}(A_y(\boldsymbol{x}), \boldsymbol{v}) = p_{ys}(z, x_y) = m_s v_y + q_s A_y(z), \tag{1.29}$$

These quantities are constants of motion in the sense that for an individual particle trajectory (the characteristics of the Vlasov equation) parameterised by t,

$$\frac{d}{dt}H_s(z(t), \boldsymbol{v}(t)) = \frac{d}{dt}p_{xs}(z(t), \boldsymbol{v}(t)) = \frac{d}{dt}p_{ys}(z(t), \boldsymbol{v}(t)) = 0,$$

where the d/dt is in fact an operator involving derivatives over phase-space,

$$\frac{d}{dt} = \frac{\partial}{\partial t} + \frac{dz}{dt}\frac{\partial}{\partial z} + \frac{d\boldsymbol{v}}{dt}\cdot\frac{\partial}{\partial \boldsymbol{v}}.$$

Using these relationships, it is now clear how one can justify writing the equilibrium DF as a function of the constants of motion

$$f_s(\boldsymbol{x}, \boldsymbol{v}) = f_s(z, \boldsymbol{v}) = f_s(H_s(z, \boldsymbol{v}), p_{xs}(z, \boldsymbol{v}), p_{ys}(z, \boldsymbol{v})),$$

and a solution of Vlasov's equation. Note how the second equality above demonstrates that the non-uniqueness of the correspondences,

$$\begin{aligned} z &= z(H_s, p_{xs}, p_{ys}), \\ \boldsymbol{v} &= \boldsymbol{v}(H_s, p_{xs}, p_{ys}), \end{aligned} \tag{1.30}$$

could play a role in this problem, see e.g. Grad (1961), Belmont et al. (2012) for discussions of this problem.

In order to now satisfy the equilibrium Maxwell equations, scalar and vector potentials must be found that satisfy the following,

$$\begin{aligned} -\epsilon_0 \frac{d^2}{dz^2}\phi(z) &= \sigma(\phi(z), A_x(z), A_y(z)) = \sum_s \int f_s(H_s, p_{xs}, p_{ys})\, d^3v, \\ -\frac{1}{\mu_0}\frac{d^2}{dz^2}A_x(z) &= j_x(\phi(z), A_x(z), A_y(z)) = \sum_s \int v_x\, f_s(H_s, p_{xs}, p_{ys})\, d^3v, \\ -\frac{1}{\mu_0}\frac{d^2}{dz^2}A_y(z) &= j_y(\phi(z), A_x(z), A_y(z)) = \sum_s \int v_y\, f_s(H_s, p_{xs}, p_{ys})\, d^3v. \end{aligned}$$

Since the RHS of the above equations are in principle now known functions of (ϕ, A_x, A_y), the problem of finding a VM equilibrium has been reduced to solving 3 coupled (ordinary) differential equations, subject to boundary conditions, e.g. the asymptotic values of the potentials at $z = \pm\infty$.

1.3.4.1 A Route Through the Forward Problem

To demonstrate how the forward problem works, we give an example for a form of DF that could be used,

$$f_s = \frac{n_{0s}}{(\sqrt{2\pi}v_{\mathrm{th},s})^3} e^{-\beta_s H_s}(a_s e^{\beta_s u_{xs} p_{xs}} + b_s e^{\beta_s u_{ys} p_{ys}})$$

for the constants a_s and b_s. This form is chosen as it is directly relatable to those considered in e.g. Harris (1962) and Schindler (2007), and has properties like that discussed in Sect. 1.3.2.2. With this form of DF, the charges and current densities become

$$\sigma = -\epsilon_0 \frac{d^2\phi}{dz^2} = \sum_s q_s e^{-q_s\beta_s\phi}\left[n_{as}e^{q_s\beta_s u_{xs}A_x} + n_{bs}e^{q_s\beta_s u_{ys}A_y}\right], \quad (1.31)$$

$$j_x = -\frac{1}{\mu_0}\frac{d^2A_x}{dz^2} = \sum_s q_s n_{as} u_{xs} e^{-q_s\beta_s(\phi - u_{xs}A_x)}, \quad (1.32)$$

$$j_y = -\frac{1}{\mu_0}\frac{d^2A_y}{dz^2} = \sum_s q_s n_{bs} u_{ys} e^{-q_s\beta_s(\phi - u_{ys}A_y)}, \quad (1.33)$$

for $n_{as} = n_{0s}a_s \exp(u_{xs}^2/(2v_{\text{th},s}^2))$ and $n_{bs} = n_{0s}b_s \exp(u_{ys}^2/(2v_{\text{th},s}^2))$. If we now make the assumption of quasineutrality—on the level of $\sigma(\phi, A_x, A_y) = 0$—then from consideration of Eq. (1.31), we see that one possible solution for $\phi = \phi(A_x, A_y)$ is as

$$\phi(A_x, A_y) = \frac{1}{\beta_e + \beta_i}(\beta_i u_{xi} + \beta_e u_{xe})A_x + \text{const.} = \frac{1}{\beta_e + \beta_i}(\beta_i u_{yi} + \beta_e u_{ye})A_y + \text{const.}, \quad (1.34)$$

when

$$\beta_i u_{xi} A_x = \beta_i u_{yi} A_y,$$
$$\beta_e u_{xe} A_x = \beta_e u_{ye} A_y.$$

Upon substituting Eq. (1.34) into Eqs. (1.32) and (1.33), the problem has now been reduced to solving two second order nonlinear ODEs in A_x and A_y,

$$(j_x =) - \frac{1}{\mu_0}\frac{d^2A_x}{dz^2} = j_{x0}e^{\alpha_x A_x},$$
$$(j_y =) - \frac{1}{\mu_0}\frac{d^2A_y}{dz^2} = j_{y0}e^{\alpha_y A_y},$$

for constants α_x, α_y, j_{x0} and j_{y0}. For examples/discussions of solutions to ODEs such as these, see Harris (1962), Schindler (2007), Tassi et al. (2008), Vasko et al. (2013). Note that Harris treats a problem like this in 1D, but with only one current density component; Schindler treats a 2D problem with only one current density component; Tassi treats a 2D problem in an MHD context and exploiting Lie Point symmetries, but with some 1D solutions; and Vasko also treats the 2D problem with a group theory approach, and only one current density component.

1.3.5 Inverse Approach for 1D VM Equilibria

As demonstrated by the above example, the 'forward approach' necessarily restricts the choice of electromagnetic fields that one can describe in a VM equilibrium, by the solution of differential equations. The inverse approach bypasses this restriction, since it begins with the prescription of the (electro-)magnetic fields themselves. The counterpoint to this—since the calculation of charge and current densities involves definite integration and hence a loss of information—is that there are in principle an infinite number of possible VM equilibrium DFs for a given macroscopic fluid equilibrium, e.g. see Wilson and Neukirch (2011) for an explicit demonstration of this feature.

The inverse approach is used in Alpers (1969), Channell (1976), Greene (1993), and Harrison and Neukirch (2009a) to obtain analytical solutions of VM equilibria, and in Mynick et al. (1979), Belmont et al. (2012) for numerical ones. All of these works consider 1D Cartesian coordinates, which are very frequently used in the study of waves, instabilities and reconnection (e.g. see Schindler 2007). In this work, and without loss of generality, z is taken to be the spatial coordinate on which the system depends, and so $\nabla = (0, 0, \partial/\partial z)$. Thus the particle Hamiltonian, H_s, and two of the canonical momenta p_{xs} and p_{ys} are conserved, see Eqs. (1.27–1.29).

1.3.5.1 Existence of a Vlasov Equilibrium

Resembling discussions in e.g. Bertotti (1963), Channell (1976), Mynick et al. (1979), Greene (1993), Schindler (2007), Harrison and Neukirch (2009b), we now consider the theory that describes macroscopic equilibria in one dimension, given the existence of a Vlasov equilibrium DF. The first velocity moment of the Vlasov equation in Cartesian coordinates

$$\int_{-\infty}^{\infty} \boldsymbol{v} \left(\boldsymbol{v} \cdot \frac{\partial f_s}{\partial \boldsymbol{x}} + \frac{q_s}{m_s}(\boldsymbol{E} + \boldsymbol{v} \times \boldsymbol{B}) \cdot \frac{\partial f_s}{\partial \boldsymbol{v}} \right) d^3v = \boldsymbol{0},$$

will, after a little algebra, yield the macroscopic/fluid equation of motion

$$\nabla \cdot \boldsymbol{P} = \sigma \boldsymbol{E} + \boldsymbol{j} \times \boldsymbol{B}.$$

In our 1D equilibrium geometry $B_z = j_z = E_x = E_y = 0$ automatically, for

$$\boldsymbol{B} = \nabla \times \boldsymbol{A}, \quad \boldsymbol{E} = -\nabla \phi,$$

and so this implies that force-balance is maintained by

$$\begin{aligned} \frac{d}{dz} P_{zx} &= 0, \\ \frac{d}{dz} P_{zy} &= 0, \\ \frac{d}{dz} P_{zz} &= \sigma E_z + j_x B_y - j_y B_x. \end{aligned} \tag{1.35}$$

We note here that this type of equilibrium is known as a *tangential equilibrium* (e.g. see Mottez 2004), and is characterised by

$$\boldsymbol{B}\cdot\nabla=0,\quad \boldsymbol{V}_s\cdot\nabla=0,$$

i.e. the magnetic field and bulk plasma flows are normal to the gradient direction(s).

If we now consider the dynamic component of the pressure tensor,

$$P_{zz}=\sum_s m_s\int_{-\infty}^{\infty} v_z^2\, f_s(H_s(\boldsymbol{v}^2,\phi),\, p_{xs}(v_x,A_x),\, p_{ys}(v_y,A_y))\,d^3v,$$

then we see that $P_{zz}=P_{zz}(\phi,A_x,A_y)$. Note that the pressure tensor is usually found by taking moments by $w_{zs}=v_z-V_{zs}$. But since f_s is only a function of v_z through H_s (which is a function of v_z^2), then automatically the v_z moment of f_s is zero, and so the bulk flow $V_{zs}=0$, giving $w_{zs}=v_z$. Using this knowledge of the form of P_{zz} gives

$$\frac{d}{dz}P_{zz}=\frac{d\phi}{dz}\frac{\partial P_{zz}}{\partial\phi}+\frac{dA_x}{dz}\frac{\partial P_{zz}}{\partial A_x}+\frac{dA_y}{dz}\frac{\partial P_{zz}}{\partial A_y}, \tag{1.36}$$

by the chain rule. A term-by-term comparison of Eq. (1.35) with Eq. (1.36) yields

$$\sigma=-\frac{\partial P_{zz}}{\partial\phi},$$
$$j_x=\frac{\partial P_{zz}}{\partial A_x},$$
$$j_y=\frac{\partial P_{zz}}{\partial A_y},$$

and so we see that the existence of a Vlasov equilibrium implies the existence of a *potential* function, P_{zz}, from which the charge and current densities can be calculated.

The above equations demonstrate that a reasonable first step in an attempt to find a VM equilibrium DF self-consistent with a given set of electromagnetic fields is to first find a P_{zz} function that is compatible. For example, in the case of a *force-free field* for which $\boldsymbol{j}\times\boldsymbol{B}=\boldsymbol{0}$, there is a simple procedure one can follow to calculate an expression for $P_{zz}(A_x,A_y)$ (for details relevant to force-free fields, see e.g. Harrison and Neukirch (2009b) and Chap. 3).

1.3.5.2 Equilibrium DF

The Vlasov equation can be solved by any differentiable function $f_s(H_s,p_{xs},p_{ys})$, with the additional 'physical' constraints being that f_s is also normalisable, non-negative and has velocity moments of arbitrary order (Schindler 2007). In line with numerous previous works in 1D (Sestero 1967; Alpers 1969; Channell 1976; Harrison and Neukirch 2009a; Abraham-Shrauner 2013), the work in this thesis on VM equilibria in Cartesian coordinates (Chaps. 2, 3 and 4) shall consider DFs of the form

$$f_s = \frac{n_{0s}}{(\sqrt{2\pi}v_{\text{th},s})^3}e^{-\beta_s H_s}g_s(p_{xs}, p_{ys}), \tag{1.37}$$

for g_s an as yet unknown function, to be determined. This form is chosen for the DF for the following reasons:

Integrability: $e^{-\beta_s H_s}$ scales like $e^{-v^2/(2v_{\text{th},s}^2)}$, implying that for a reasonable g_s function, all moments of f_s will be integrable, as necessary,

Solving integrals: The $e^{-v^2/(2v_{\text{th},s}^2)}$ dependence lends itself to not only being integrable, but to having known definite integrals when multiplied by many functions,

Physical meaning: As discussed in Sect. 1.2.1, the unique equilibrium DF for a collisional plasma is a Maxwellian. As such it is clear how this Vlasov/collisionless equilibrium DF relates to a collsional equilibrium DF,

Elegance: A zero-flow Maxwellian DF is reproduced when $g_s = 1$.

1.3.5.3 Scalar and Vector Potentials

As demonstrated in Sect. 1.3.4.1, the combination of quasineutrality,

$$n_i(A_x, A_y, \phi) = n_e(A_x, A_y, \phi),$$

and a DF of the form in Eq. (1.37) results in a scalar potential that is implicitly defined as a function of the vector potential, e.g. Harrison and Neukirch (2009b, Schindler (2007), Tasso and Throumoulopoulos (2014), and Nakariakov (2015):

$$\phi_{qn}(A_x, A_y) = \frac{1}{e(\beta_e + \beta_i)}\ln(n_i/n_e). \tag{1.38}$$

In Chaps. 2, 3 and 4, and as in e.g. Channell (1976), parameters will be chosen such that $n_i = n_e$ as functions over (A_x, A_y) space, and so 'strict neutrality' is satisfied, implying $\phi_{qn} = 0$. This choice of parameters is mathematically equivalent to the condition used to derive the 'micro-macroscopic' parameter relationships, which will be discussed later.

It has been commented in e.g. Grad (1961), Bertotti (1963), Nicholson (1963), Sestero (1966), Mynick et al. (1979), Attico and Pegoraro (1999), and Harrison and Neukirch (2009b), that the 1D VM equilibrium problem is analagous to that of a particle moving under the influence of a potential; with the relevant component of the pressure tensor, P_{zz}, taking the role of the potential; (A_x, A_y) the role of position and z the role of time. This analogy is demonstrated by

$$\frac{d^2 A_x}{dz^2} = -\mu_0 \frac{\partial P_{zz}}{\partial A_x}, \tag{1.39}$$

$$\frac{d^2 A_y}{dz^2} = -\mu_0 \frac{\partial P_{zz}}{\partial A_y}. \tag{1.40}$$

The LHS of the above equations take the role of acceleration, and the RHS take the role of force, as the gradient of a potential. Through this analogy, the task of finding a consistent P_{zz} function—as discussed in Sect. 1.3.5.1—can be reformulated as finding a 'potential function' P_{zz}, such that a 'particle trajectory' follows $(A_x(z), A_y(z))$.

1.3.5.4 The Inverse Problem

Channell (1976) developed the theory of the inverse problem in a general sense, with the assumption of zero scalar potential from the offset. It is shown therein that a DF of the form of Eq. (1.37) implies that the relevant component of the pressure tensor, P_{zz}, is a 2-D integral transform of the unknown function g_s, given by

$$P_{zz}(A_x, A_y) = \frac{\beta_e + \beta_i}{\beta_e \beta_i} \frac{n_{0s}}{2\pi m_s^2 v_{\text{th},s}^2} \times \int_{-\infty}^{\infty} \int_{-\infty}^{\infty} e^{-\beta_s \left((p_{xs} - q_s A_x)^2 + (p_{ys} - q_s A_y)^2\right)/(2m_s)} g_s(p_{xs}, p_{ys}) dp_{xs} dp_{ys}. \tag{1.41}$$

This equation together with Eqs. (1.39) and (1.40) define the inverse problem at hand, viz. 'for a given macroscopic equilibrium described by $(A_x(z), A_y(z))$, can we find a self-consistent $P_{zz}(A_x, A_y)$ according to Eqs. (1.39) and (1.40), and can we then invert the integral transform in Eq. (1.41) to solve for the unknown function g_s?' Observe that the LHS of Eq. (1.41) is species-independent, whereas the RHS seems not to be. In fact, the consistency of this equation for both ions and electrons is one more condition that is implicit in 'Channell's method', and is formally compatible with the condition of strict neutrality, $\phi = 0$.

1.3.5.5 Inversion by Fourier Transforms

As written, Eq. (1.41) is almost exactly a 2D *convolution* of the functions $e^{-(t_1^2+t_2^2)/2}$ and $g(t_1, t_2)$, for a convolution of functions $h_1(t_1, t_2)$ and $h_2(t_1, t_2)$ defined as

$$\begin{aligned} h_1 \star h_2 (\tau_1, \tau_2) &= \int_{t_1=-\infty}^{t_1=\infty} \int_{t_2=-\infty}^{t_2=\infty} h_1(\tau_1 - t_1, \tau_2 - t_2) h_2(t_1, t_2) dt_2 dt_1, \\ &= \int_{t_1=-\infty}^{t_1=\infty} \int_{t_2=-\infty}^{t_2=\infty} h_1(t_1, t_2) h_2(\tau_1 - t_1, \tau_2 - t_2) dt_2 dt_1. \end{aligned} \tag{1.42}$$

There is a useful result regarding the Fourier transform,

$$\mathrm{FT}[h](\omega) = \frac{1}{\sqrt{2\pi}} \int_{-\infty}^{\infty} e^{-it\omega} h(t) dt,$$

of a convolution. The convolution theorem states that

$$\mathrm{FT}[h_1 \star h_2](\omega_1, \omega_2) = \mathrm{FT}[h_1](\omega_1)\,\mathrm{FT}[h_2](\omega_2),$$

(Zayed 1996). That is to say that the Fourier transform of a convolution of functions is the product of the transforms of the individual functions. By making some simple changes of variables, $\boldsymbol{\mathcal{A}} = \boldsymbol{A}/q_s$, Eq. (1.41) can be manipulated into the form of Eq. (1.42),

$$P_{zz}\left(\frac{A_x}{q_s}, \frac{A_y}{q_s}\right) = P_{zz}(\mathcal{A}_{xs}, \mathcal{A}_{ys}) = \frac{\beta_e + \beta_i}{\beta_e \beta_i} \frac{n_{0s}}{2\pi m_s^2 v_{\mathrm{th},s}^2} e^{-\beta_s (p_{xs}^2 + p_{ys}^2)/(2m_s)} \star g_s. \tag{1.43}$$

As such, and using the convolution theorem, g_s can—at least formally—be written

$$g_s(p_{xs}, p_{ys}) = \frac{\beta_e \beta_i}{\beta_e + \beta_i} \frac{2\pi m_s^2 v_{\mathrm{th},s}^2}{n_{0s}} \mathrm{IFT}\left[\frac{\mathrm{FT}[P_{zz}](\omega_1, \omega_2)}{\mathrm{FT}\left[e^{-\beta_s (t_1^2 + t_2^2)/(2m_s)}\right](\omega_1, \omega_2)}\right], \tag{1.44}$$

for IFT the inverse Fourier transform,

$$\mathrm{IFT}[h](t) = \frac{1}{\sqrt{2\pi}} \int_{-\infty}^{\infty} e^{it\omega} \mathrm{FT}[h](\omega) d\omega.$$

Note that the $t_1, t_2, \omega_1, \omega_2$ variables used in Eq. (1.44) are in a sense dummy variables, and do not in fact represent time/frequency in this example, but were used for consistency with the rest of the discussion. For dimensional consistency the conjugate variables to the p_{xs}, p_{ys} variables should have dimensions of "1/momentum".

This Fourier transform method is used in Channell (1976); Harrison and Neukirch (2009a) to derive VM equilibrium DFs for 1D macroscopic equilibria, and in a sense this is the most natural method for the problem. At least, one can always formally write down the solution. However, there are two main difficulties:

Integrability: Since the Fourier transform of a Gaussian is a Gaussian (Erdelyi et al. 1954), part of the RHS of Eq. (1.44) is an exponential of a positive quadratic. Formally, the integrability of the RHS places serious restrictions on the nature of $\mathrm{FT}[P_{zz} : (\omega_1, \omega_2)]$, and hence the validity of the method. We note here that despite this formal restriction on the use of the Fourier transform, it is in effect possible to bypass this problem by inspection. For example, in Neukirch et al. (2009), Abraham-Shrauner (2013) the g_s function is found 'by inspection'/using known integrals, that give the same result that the (invalid) Fourier transform method would have.

Integrals: It may be that certain P_{zz} functions in Eq. (1.43) have no analytic expression for the Fourier transform, or that the argument of the RHS of Eq. (1.44) has no analytic expression for the inverse Fourier transform.

1.3.6 Previous Work on VM Equilibria

In this thesis we shall consider theory and examples of exact self-consistent solutions of the VM system for magnetised plasmas, including some non-trivial solutions of Poisson's equation such that the plasma can be either neutral or non-neutral, in Chap. 5. Our focus will be on translationally invariant equilibria in Cartesian geometry in Chaps. 2, 3 and 4, and on rotationally symmetric equilibria in cylindrical geometry in Chap. 5. These solutions can either describe equilibria of the VM system, such that the one-particle DF for species s, f_s, satisfies the steady-state Vlasov equation in particle phase space $(\boldsymbol{x}, \boldsymbol{v})$,

$$\frac{df_s(\boldsymbol{x}, \boldsymbol{v}; t)}{dt} = 0 = \frac{\partial f_s(\boldsymbol{x}, \boldsymbol{v}; t)}{\partial t},$$

or as aforementioned in Sect. 1.3, nonlinear wave solutions that satisfy the above equation when Galilei-transformed to the wave frame (e.g. see Bernstein et al. 1957; Abraham-Shrauner 1968), by making a transformation

$$\begin{aligned}\boldsymbol{x} &\to \boldsymbol{x} - \boldsymbol{u}t, \\ \boldsymbol{v} &\to \boldsymbol{v} - \boldsymbol{u},\end{aligned}$$

for $\boldsymbol{u}$ the phase velocity of the travelling wave.

Knowledge of exact solutions to the VM system are of value in the study of a wide variety of phenomena in collisionless plasmas, and a comprehensive review and description of all the potential applications is beyond the scope of this thesis. However, we shall survey the theoretical works most relevant to ours, and some applications. Broadly speaking there are three approaches in the literature: on electrostatic and un-magnetised; electrostatic and magnetised; and neutral magnetised plasmas. Of course these 'streams' have some overlap, and theoretically the boundary between them is 'woolly' by the Lorentz invariance of Maxwell's equations. Specifically, since Galilean frame transformations, $\boldsymbol{u}$—in the non-relativistic scenario where $u \ll c$—can send

$$\boldsymbol{E}' = \boldsymbol{0} \to \boldsymbol{E} = \boldsymbol{u} \times \boldsymbol{B}, \quad \text{or} \tag{1.45}$$

$$\boldsymbol{B}' = \boldsymbol{0} \to \boldsymbol{B} = -\frac{1}{c^2}\boldsymbol{u} \times \boldsymbol{E}, \tag{1.46}$$

(e.g. see Griffiths 2013; Landau and Lifshitz 2013). We interpret Eqs. (1.45) and (1.46) as follows. Consider two coordinate systems: the stationary laboratory, K, and one moving at a constant velocity $\boldsymbol{u}$ relative to the laboratory, K'. In these two coordinate systems, the electromagnetic fields are denoted without and with primes, respectively. Then Eq. (1.45) says that if in the frame K' the electric field is measured to be $\boldsymbol{E}' = \boldsymbol{0}$, then it measured to be given by $\boldsymbol{u} \times \boldsymbol{B}$ in the K frame. Likewise, Eq. (1.46) says that if in the frame K' the magnetic field is measured to be $\boldsymbol{B}' = \boldsymbol{0}$, then it measured to be given by $c^{-2}\boldsymbol{u} \times \boldsymbol{E}$ in the K frame.

Not only that, but the differences/distinctions between the following frequently assumed states:

- 'strict neutrality' (e.g. see Grad 1961; Channell 1976),

$$\phi = 0 \implies \sigma = 0;$$

- quasineutrality, i.e. $\sigma = 0$ to first order, as introduced in Sect. 1.1.3), and typically achieved in the literature (e.g. see Harrison and Neukirch 2009b) by

$$\phi = \phi(\boldsymbol{A}(\boldsymbol{x})) \quad \text{s.t.} \quad \sigma = 0;$$

- non-neutrality (e.g. see Davidson 2001 for the authoritative text),

$$\phi = \phi(\boldsymbol{x}) \quad \text{s.t.} \quad \sigma \neq 0,$$

are subtle (e.g. see Bertotti 1963; Greene 1993; Schindler 2007). Given these considerations, we shall make some crude distinctions, and given that the electrostatic literature is relatively self-contained and seemingly the one that gained maturity the quickest, we describe this first.

The seminal work on electrostatic solutions of the VM system in the absence of a magnetic field is that of Bernstein et al. (1957), in which an inductive method is developed that calculates the DF of trapped electrons in a nonlinear travelling electrostatic wave (*Bernstein-Greene-Kruskal (BGK) waves*), for a given 1D scalar potential, ϕ, in the wave frame. This work was developed upon in particular by Schamel (1971, 1972a) with particular emphasis on the necessary condition of positivity of the DF. Other theoretical works in a 1D geometry include those on ion-acoustic waves (e.g. see Schamel 1972b, ion/electron holes and double layers (e.g. see Schamel 1986, 2000), generalisations and extensions of BGK theory (e.g. see Lewis and Symon 1984; Karimov and Lewis 1999), and 'three-dimensional BGK waves' (e.g. see Ng and Bhattacharjee 2005; Ng et al. 2006). One particular application of this theory is the phenomena of collisionless shocks (e.g. see Burgess and Scholer 2015; Marcowith et al. 2016), relevant in astrophysical, laboratory, and laboratory astrophysical contexts (e.g. see Montgomery and Joyce 1969; Forslund and Shonk 1970; Forslund and Freidberg 1971; Eliasson and Shukla 2006; Spitkovsky 2008; Stockem et al. 2014; Cairns et al. 2014; Svedung Wettervik et al. 2016).

There exists a similarly rich literature for magnetised quasi-neutral and non-neutral solutions (the majority of which is quasi-neutral), much of which is collected in the articles by Roth et al. (1996), Zelenyi et al. (2011), and Artemyev and Zelenyi

(2013). Perhaps the most ubiquitous work in the context of current sheets is that of Harris (1962), in which it is demonstrated that the DF consistent with the 1D Harris current sheet and for a plasma with zero scalar potential can, by using a post-hoc Galilean transformation, also describe a non-neutral configuration (the Harris sheet equilibrium is considered in the relativistic case in Hoh 1966). The foundational work in the realm of magnetised and electrostatic collisionless shocks is that of Sagdeev (1966), in which analogies are drawn between the equations describing solitary waves, and the motion of a particle in a potential: the Sagdeev potential. General theoretical treatments on quasi-neutral and non-neutral VM equilibria include, for

- 1D plasmas: Tonks (1959), Sestero (1964, 1966, 1967), Lam (1967), Abraham-Shrauner (1968), Lemaire and Burlaga (1976), Lee and Kan (1979), Mitchell and Kan (1979), Greene (1993), Mottez (2003), Yoon et al. (2006), Balikhin and Gedalin (2008), Artemyev (2011),
- Two-dimensional (2D) plasmas: Hewett et al. (1976), Mynick et al. (1979), Kan (1979), Otto and Schindler (1984), Muschietti et al. (2000), Schindler and Birn (2002), Eliasson et al. (2006), Suzuki and Shigeyama (2008), Kocharovsky et al. (2010), Schindler (2007), Ng et al. (2012), and Vasko et al. (2013),
- With applications to magnetospheres for 1D plasmas: Davies (1968), (1969), Su and Sonnerup (1971), Kan and Akasofu (1979), Stern (1981a, b), Rogers and Whipple (1988), and DeVore et al. (2015),
- With applications to magnetospheres for 2D plasmas: Kan et al. (1979), Lee and Kan (1979a), Birn et al. (2004).

For theoretical treatments that treat the plasma as strictly neutral ($\phi = 0$), see Grad (1961, Hurley (1963), Nicholson (1963), Schmid-Burgk (1965), Moratz and Richter (1966), Lerche (1967), Alpers (1969), Channell (1976), Bobrova and Syrovatskii (1979), Lakhina and Schindler (1983), Attico and Pegoraro (1999), Bobrova et al. (2001), Fu and Hau (2005), Yoon and Lui (2005), Harrison and Neukirch (2009a), Neukirch et al. (2009), Panov et al. (2011), Wilson and Neukirch (2011), Belmont et al. (2012), Janaki and Dasgupta (2012), Abraham-Shrauner (2013), Ghosh et al. (2014), Kolotkov et al. (2015), Allanson et al. (2015), and Allanson et al. (2016).

We should indicate that there also exists a substantial literature on magnetised neutral and non-neutral VM solutions in cylindrical geometry (for example flux tubes, mono-energetic beams, laboratory pinches and astrophysical jets), with Davidson (2001), Vinogradov et al. (2016), and Allanson et al. (2016) and references therein, as well as Chap. 5 providing a suitable starting point for an interested reader.

1.4 Thesis Motivation and Outline

The importance of understanding the equilibrium states permitted by a given system is common to most physical disciplines, and this is—broadly speaking—the motivation for the work in this thesis. Specifically, I shall consider electromagnetic structures that—by the balance of electromagnetic, inertial, and thermal pressure forces—confine the mass and electric currents in a plasma. These equilibrium configurations will be considered in Cartesian and cylindrical geometries, namely current sheets and flux tubes. There are many potential applications for current sheet and flux tube equilibria, and these shall be discussed in Chaps. 3, 4, and then 5 respectively. However, the main/most timely application of the work in this thesis could be to studies of magnetic reconnection, for which localised currents are a pre-condition.

1.4.1 Outline of the Thesis

This thesis is structured as follows:

- Chapter 2: **The Use of Hermite Polynomials for the Inverse Problem in Nne-Dimensional Vlasov-Maxwell Equilibria**
  By expressing the unknown functions, g_s, of the canonical momenta as (infinite) expansions of Hermite polynomials, we establish a one-to-one correspondence between the coefficients of expansion, and those of a Maclaurin expansion of the pressure tensor. We then find a sufficient condition for the convergence of the Hermite representation, contingent on the Maclaurin expansion coefficients of the pressure tensor. For certain classes of DFs, we prove results on the non-negativity of the g_s function, and make a conjecture for all other classes.
- Chapter 3: **One-Dimensional Nonlinear Force-Free Current Sheets**
  Using pressure transformation techniques, we find a new pressure tensor self-consistent with the force-free Harris sheet magnetic field, for any value of the plasma beta, and crucially sub-unity values that could not be accessed before. Then we use the Hermite polynomial expansion technique established in Chap. 2 to calculate a Vlasov equilibrium DF consistent with the low beta force-free Harris sheet. Next, the Hermite expansion is proven to be analytically convergent, using the sufficient condition derived in Chap. 2, and we confirm that the DF satisfies the conjectured condition for non-negativity of the Hermite representation of a DF, also from Chap. 2.

  We conduct a preliminary analysis on the physical properties of the DF, but encounter numerical difficulties for the parameter range of interest when attempting to make plots for $\beta_{pl} < 0.85$. In response to this difficulty, we 're-gauge' the vector potential, allowing for numerical convergence of the Hermite expansions for much lower values of the plasma beta, $\beta_{pl} = 0.05$. As before, we establish the necessary convergence and non-negativity of the DF, and present new plots.
- Chapter 4: **One-Dimensional Asymmetric Current Sheets**
  We first consider the mathematical problem for a pressure tensor consistent with an 'asymmetric' current sheet equilibrium. Using these results, we present possible examples of pressure tensors self-consistent with asymmetric equilibria, and discuss the inverse problem. It becomes apparent that for certain representations,

the problem is not analytically soluble, and numerical techniques are necessary. Using representations for the pressure tensor that give soluble solutions, we present exact analytic VM equilibria for an asymmetric Harris sheet with guide field, and a preliminary analysis

- Chapter 5: **Neutral and Non-neutral Flux Tube Equilibria**
  This is a departure from the previous work on translationally invariant systems. First we consider the problem of constructing one-dimensional VM equilibria in cylindrical geometry, and establish the fluid equation(s) of motion from the Vlasov equation in cylindrical geometry. We include an analysis of the microscopic origin of the macroscopic forces in the resultant equation of motion.
  Next, there is discussion on the attempts to construct VM equilibria for the exact Gold-Hoyle model, a force-free flux tube. These attempts do not yield solutions, and there seems to be good physical reasoning behind the mathematical difficulties. By making a small change to the macroscopic magnetic field, we are able to find a consistent VM equilibrium for the Gold-Hoyle model embedded in a uniform background field. We present a preliminary analysis of the equilibrium, including a consideration of multiple maxima in velocity space, and the non-neutrality of the macroscopic configuration.
- Chapter 6: **Discussion**
  We briefly summarise the main results from this thesis, and place them in the context of current plasma physics research. In particular we focus on open questions and avenues that merit further investigation.

References

B. Abraham-Shrauner, Exact, stationary wave solutions of the nonlinear Vlasov equation. Phys. Fluids **11**, 1162–1167 (1968)

B. Abraham-Shrauner, Force-free Jacobian equilibria for Vlasov-Maxwell plasmas. Phys. Plasmas **20**(10), 102117 (2013)

O. Allanson, T. Neukirch, S. Troscheit, F. Wilson, From onedimensional fields to Vlasov equilibria: theory and application of Hermite polynomials. J. Plasma Phys. 82.3, 905820306 (2016)

O. Allanson, T. Neukirch, F. Wilson, S. Troscheit, An exact collisionless equilibrium for the Force-Free Harris Sheet with low plasma beta. Phys. Plasmas 22.10, 102116 (2015)

O. Allanson, F. Wilson, T. Neukirch, Neutral and non-neutral collisionless plasma equilibria for twisted flux tubes: The Gold-Hoyle model in a background field. Phys. Plasmas **23**(9), 092106 (2016)

W. Alpers, Steady state charge neutral models of the magnetopause. Astrophys. Space Sci. **5**, 425–437 (1969)

A.V. Artemyev, A model of one-dimensional current sheet with parallel currents and normal component of magnetic field. Phys. Plasmas **18**(2), 022104 (2011)

A. Artemyev, L. Zelenyi, Kinetic structure of current sheets in the earth magnetotail. Space Sci. Rev. **178**, 419–440 (2013)

N. Attico, F. Pegoraro, Periodic equilibria of the Vlasov-Maxwell system. Phys. Plasmas **6**, 767–770 (1999)

N. Aunai, M. Hesse, S. Zenitani, M. Kuznetsova, C. Black, R. Evans, R. Smets, Comparison between hybrid and fully kinetic models of asymmetric magnetic reconnection: coplanar and guide field configurations. Phys. Plasmas 20.2, 022902, 022902 (2013)

M. Balikhin, M. Gedalin, Generalization of the Harris current sheet model for non-relativistic, relativistic and pair plasmas. J. Plasma Phys. **74**, 749–763 (2008)

W. Baumjohann, R.A. Treumann, *Basic Space Plasma Physics*. (Imperial College Press, 1997)

M.T. Beidler, P.A. Cassak, Model for incomplete reconnection in Sawtooth crashes. Phys. Rev. Lett. 107.25, 255002, 255002 (2011)

G. Belmont, N. Aunai, R. Smets, Kinetic equilibrium for an asymmetric tangential layer. Phys. Plasmas **19**(2), 022108 (2012)

I.B. Bernstein, J.M. Greene, M.D. Kruskal, Exact nonlinear plasma oscillations. Phys. Rev. **108**, 546–550 (1957)

B. Bertotti, Fine structure in current sheaths. Ann. Phys. **25**, 271–289 (1963)

J. Birn, K. Galsgaard, M. Hesse, M. Hoshino, J. Huba, G. Lapenta, P. L. Pritchett, K. Schindler, L. Yin, J. B?chner, T. Neukirch, E.R. Priest, Forced magnetic reconnection. Geophys. Res. Lett. 32.6, L06105 (2005)

J. Birn, M. Hesse, Energy release and transfer in guide field reconnection. Phys. Plasmas 17.1, 012109, 012109 (2010)

J. Birn, E. Priest, *Reconnection of Magnetic Fields: Magnetohydrodynamics and Collisionless Theory and Observations*. (Cambridge University Press, 2007)

J. Birn, J.F. Drake, M.A. Shay, B.N. Rogers, R.E. Denton, M. Hesse, M. Kuznetsova, Z.W. Ma, A. Bhattacharjee, A. Otto, P.L. Pritchett, Geospace environmental modeling (GEM) magnetic reconnection challenge. J. Geophys. Res. Space Phys. **106**(A3), 3715–3719 (2001)

J. Birn, K. Schindler, M. Hesse, Thin electron current sheets and their relation to auroral potentials. J. Geophys. Res. Space Phys. **109**(A2), A02217 (2004)

D. Biskamp, Magnetic reconnection in plasmas. Magnetic reconnection in plasmas, Cambridge, UK: Cambridge University Press, 2000 xiv, 387 p. Cambridge monographs on plasma physics, vol. 3, ISBN 0521582881 (2000)

N.A. Bobrova, S.V. Bulanov, J.I. Sakai, D. Sugiyama, Force-free equilibria and reconnection of the magnetic field lines in collisionless plasma configurations. Phys. Plasmas **8**, 759–768 (2001)

N.A. Bobrova, S.I. Syrovatskii, Violent instability of one-dimensional forceless magnetic field in a rarefied plasma. Sov. J. Exp. Theor. Phys. Lett. **30**, 535-+ (1979)

A. Borissov, T. Neukirch, J. Threlfall, Particle acceleration in collapsing magnetic traps with a braking plasma jet. Solar Phys. **291**, 1385–1404 (2016)

S.I. Braginskii, Transport processes in a plasma. Rev. Plasma Phys. **1**, 205 (1965)

S.G. Brush, The kinetic theory of gases. World Sci., 262–349 (2003)

D. Burgess, M. Scholer, *Collisionless Shocks in Space Plasmas: Structure and Accelerated Particles* (Cambridge University Press, Cambridge Atmospheric and Space Science Series, 2015)

R.A. Cairns, R. Bingham, P. Norreys, R. Trines, Laminar shocks in high power laser plasma interactions. Phys. Plasmas 21.2, 022112, 022112 (2014)

J.R. Cary, A.J. Brizard, Hamiltonian theory of guiding-center motion. Rev. Modern Phys. **81**, 693–738 (2009)

P.J. Channell, Exact Vlasov-Maxwell equilibria with sheared magnetic fields. Phys. Fluids **19**, 1541–1545 (1976)

S.C. Cowley, Lecture Notes for the UCLA course. http://plasma.physics.ox.ac.uk/plasma/Courses.html (2003/4)

R.C. Davidson, *Physics of Nonneutral Plasmas*. (World Scientific Press, 2001)

C.M. Davies, The boundary layer between a cold plasma and a confined magnetic field when the plasma is not normally incident on the boundary. Planetary Space Sci. **16**, 1249-+ (1968)

C.M. Davies, The structure of the magnetopause. Planetary Space Sci. **17**, 333-+ (1969)

C.R. DeVore, S.K. Antiochos, C.E. Black, A.K. Harding, C. Kalapotharakos, D. Kazanas, A.N. Timokhin, A model for the electrically charged current sheet of a pulsar. Astrophys. J. **801**(109), 109 (2015)

J.F. Drake, Y.C. Lee, Kinetic theory of tearing instabilities. Phys. Fluids **20**, 1341–1353 (1977)
J.W. Dungey, Interplanetary magnetic field and the auroral zones. Phys. Rev. Lett. **6**, 47–48 (1961)
B. Eliasson, P.K. Shukla, Formation and dynamics of relativistic electromagnetic solitons in plasmas containing high-and low-energy electron components. Sov. J. Exp. Theor. Phys. Lett. **83**, 447–452 (2006)
B. Eliasson, P.K. Shukla, M.E. Dieckmann, Theory and simulations of nonlinear kinetic structures in plasmas. Plasma Phys. Controll. Fusion **48**, B257–B265 (2006)
A. Erdelyi, W. Magnus, F. Oberhettinger, F.G. Tricomi, *Tables of integral transforms*, vol. I. Based, in part, on notes left by Harry Bateman. (McGraw- Hill Book Company, Inc., New York-Toronto-London, 1954), pp. xx+391
D.F. Escande, F. Doveil, Y. Elskens, N-body description of Debye shielding and Landau damping. Plasma Phys. Controll. Fusion 58.1, 014040, 014040 (2016)
R. Fitzpatrick, *Plasma Physics: An Introduction* (CRC Press, Taylor & Francis Group, 2014)
D.W. Forslund, J.P. Freidberg, Theory of laminar collisionless shocks. Phys. Rev. Lett. **27**, 1189–1192 (1971)
D.W. Forslund, C.R. Shonk, Formation and structure of electrostatic Collisionless shocks. Phys. Rev. Lett. **25**, 1699–1702 (1970)
J.P. Freidberg, *Ideal Magnetohydrodynamics*. (Plenum Publishing Corportation, 1987)
G. Fruit, P. Louarn, A. Tur, D. Le QuAu, On the propagation of magnetohydrodynamic perturbations in a Harris-type current sheet 1. Propagation on discrete modes and signal reconstruction. J. Geophys. Res. (Space Physics) 107, 1411, SMP 39-1-SMP 39–18 (2002)
W.-Z. Fu, L.-N. Hau, Vlasov-Maxwell equilibrium solutions for Harris sheet magnetic field with Kappa velocity distribution. Phys. Plasmas 12.7, pp. 070701-+ (2005)
H.P. Furth, J. Killeen, M.N. Rosenbluth, Finite-resistivity instabilities of a sheet pinch. Phys. Fluids **6**, 459–484 (1963)
S.P. Gary, *Theory of Space Plasma Microinstabilities* (Cambridge University Press, Cambridge Atmospheric and Space Science Series, 2005)
A. Ghosh, M.S. Janaki, B. Dasgupta, A. Bandyopadhyay, Chaotic magnetic fields in Vlasov-Maxwell equilibria. Chaos 24.1, 013117, 013117 (2014)
H. Grad, Boundary layer between a plasma and a magnetic field. Phys. Fluids **4**, 1366–1375 (1961)
H. Grad, On the kinetic theory of rarefied gases. Comm. Pure Appl. Math. **2**, 331–407 (1949)
J.M. Greene, One-dimensional Vlasov-Maxwell equilibria. Phys. Fluids B **5**, 1715–1722 (1993)
D.J. Griffiths, *Introduction to Electrodynamics*. (Pearson, 2013)
F. Guo, H. Li, W. Daughton, Y.-H. Liu, Formation of hard power laws in the energetic particle spectra resulting from relativistic magnetic reconnection. Phys. Rev. Lett. **113**, 155005 (2014)
E.G. Harris, On a plasma sheath separating regions of oppositely directed magnetic field. Nuovo Cimento **23**, 115 (1962)
M.G. Harrison, Equilibrium and dynamics of collisionless current sheets (2009). The University of St Andrews, PhD thesis
M.G. Harrison, T. Neukirch, One-dimensional Vlasov-Maxwell equilibrium for the force-free harris sheet. Phys. Rev. Lett. 102.13, 135003-+ (2009a)
M.G. Harrison, T. Neukirch, Some remarks on one-dimensional forcefree Vlasov-Maxwell equilibria. Phys. Plasmas 16.2, 022106-+ (2009b)
M. Hesse, N. Aunai, D. Sibeck, J. Birn, On the electron diffusion region in planar, asymmetric, systems. Geophys. Res. Lett. **41**, 8673–8680 (2014)
M. Hesse, J. Birn, M. Kuznetsova, Collisionless magnetic reconnection: electron processes and transport modeling. J. Geophys. Res. **106**, 3721–3736 (2001)
M. Hesse, M. Kuznetsova, K. Schindler, J. Birn, Three-dimensional modeling of electron quasiviscous dissipation in guide-field magnetic reconnection. Phys. Plasmas 12.10, pp. 100704-+ (2005)
M. Hesse, K. Schindler, A theoretical foundation of general magnetic reconnection. J. Geophys. Res. **93**, 5559–5567 (1988)

M. Hesse, T. Neukirch, K. Schindler, M. Kuznetsova, S. Zenitani, The diffusion region in collisionless magnetic reconnection. Space Sci. Rev. **160**(1), 3–23 (2011)
M. Hesse, N. Aunai, S. Zenitani, M. Kuznetsova, J. Birn, Aspects of collisionless magnetic reconnection in asymmetric systems. Phys. Plasmas **20**(6), 061210 (2013)
M. Hesse, N. Aunai, J. Birn, P. Cassak, R.E. Denton, J.F. Drake, T. Gombosi, M. Hoshino, W. Matthaeus, D. Sibeck, S. Zenitani, Theory and modeling for the magnetospheric multiscale mission. Space Sci. Rev. **199**(1), 577–630 (2016)
D.W. Hewett, C.W. Nielson, D. Winske, Vlasov confinement equilibria in one dimension. Phys. Fluids **19**, 443–449 (1976)
F.C. Hoh, Stability of sheet pinch. Phys. Fluids **9**, 277–284 (1966)
J. Hurley, Analysis of the transition region between a uniform plasma and its confining magnetic field. II. Phys. Fluids **6**, 83–88 (1963)
I.H. Hutchinson, Electron holes in phase space: What they are and why they matter. Phys. Plasmas **24**(5), 055601 (2017)
M.S. Janaki, B. Dasgupta, Vlasov-Maxwell equilibria: examples from higher-curl Beltrami magnetic fields. Phys. Plasmas **19**(3), 032113 (2012)
J.R. Kan, Non-linear tearing structures in equilibrium current sheet. Planetary Space Sci. **27**, 351–354 (1979)
J.R. Kan, S.-I. Akasofu, A model of the auroral electric field. J. Geophys. Res. **84**, 507–512 (1979)
J.R. Kan, L.C. Lee, S.-I. Akasofu, Two-dimensional potential double layers and discrete auroras. J. Geophys. Res. **84**, 4305–4315 (1979)
A.R. Karimov, H.R. Lewis, Nonlinear solutions of the Vlasov-Poisson equations. Phys. Plasmas **6**, 759–761 (1999)
V.V. Kocharovsky, V.V. Kocharovsky, V.J. Martyanov, Self-consistent current sheets and filaments in relativistic collisionless plasma with arbitrary energy distribution of particles. Phys. Rev. Lett. 104.21, 215002, 215002 (2010)
D.Y. Kolotkov, I.Y. Vasko, V.M. Nakariakov, Kinetic model of forcefree current sheets with non-uniform temperature. Phys. Plasmas 22.11, 112902, 112902 (2015)
N.A. Krall, A.W. Trivelpiece, *Principles of plasma physics. International Student Edition—International Series in Pure and Applied Physics*. (McGraw-Hill, Tokyo Kogakusha, 1973)
R.M. Kulsrud, MHD description of plasma, in *Handbook of Plasma Physics*, vol. 1, ed. by A.A. Galeev, R.N. Sudan (North- Holland, Amsterdam, 1983)
G.S. Lakhina, K. Schindler, Tearing modes in the magnetopause current sheet. Astrophys. Space Sci. **97**, 421–426 (1983)
S.H. Lam, One-dimensional static pinch solutions. Phys. Fluids **10**, 2454–2457 (1967)
L.D. Landau, E.M. Lifshitz, *The Classical Theory of Fields* (Elsevier Science, Course of Theoretical Physics, 2013)
A. Lazarian, E.M. de Gouveia Dal Pino, C. Melioli (eds.) Magnetic fields in diffuse media, vol. 407 (Astrophysics and Space Science Library, 2015)
A. Lazarian, E.T. Vishniac, Reconnection in a weakly stochastic field. Astrophys. J. **517**, 700–718 (1999)
L.C. Lee, J.R. Kan, A unified kinetic model of the tangential magnetopause structure. J. Geophys. Res. (Space Physics) **84**, 6417–6426 (1979a)
L.C. Lee, J.R. Kan, Transition layer between two magnetized plasmas. J. Plasma Phys. **22**, 515–524 (1979b)
J. Lemaire, L.F. Burlaga, Diamagnetic boundary layers—a kinetic theory. Astrophys. Space Sci. **45**, 303–325 (1976)
I. Lerche, On the boundary layer between aWarm, Streaming plasma and a confined magnetic field. J. Geophys. Res. (Space Physics) **72**, 5295-+ (1967)
H.R. Lewis, K.R. Symon, Exact time-dependent solutions of the Vlasov-Poisson equations. Phys. Fluids **27**, 192–196 (1984)
R.G. Littlejohn, Variational principles of guiding centre motion. J. Plasma Phys. **29**, 111–125 (1983)

Y.-H. Liu, M. Hesse, Suppression of collisionless magnetic reconnection in asymmetric current sheets. Phys. Plasmas 23.6, 060704, 060704 (2016)

N.F. Loureiro, D.A. Uzdensky, Magnetic reconnection: from the Sweet??? Parker model to stochastic plasmoid chains. Plasma Phys. Controll. Fusion **58**(1), 014021 (2016)

K. Malakit, M.A. Shay, P.A. Cassak, C. Bard, Scaling of asymmetric magnetic reconnection: kinetic particle-in-cell simulations. J. Geophys. Res. (Space Physics) 115, A10223, A10223 (2010)

A. Marcowith, A. Bret, A. Bykov, M. E. Dieckman, L. O'C Drury, B. Lembge, M. Lemoine, G. Morlino, G. Murphy, G. Pelletier, I. Plotnikov, B. Reville, M. Riquelme, L. Sironi, A. Stockem Novo, The microphysics of collisionless shock waves. Reports Progress Phys. 79.4, 046901, 046901 (2016)

H.G. Mitchell Jr., J.R. Kan, Current interruption in a collisionless plasma by nonlinear electrostatic waves. Planetary Space Sci. **27**, 933–937 (1979)

D. Montgomery, G. Joyce, Shock-like solutions of the electrostatic Vlasov equation. J. Plasma Phys. **3**, 1–11 (1969)

T.E. Moore, J.L. Burch, R.B. Torbert, Magnetic reconnection. Nat. Phys. **11**, 611–613 (2015)

E. Moratz, E.W. Richter, Elektronen-Geschwindigkeitsverteilungsfunktionen fr kraftfreie bzw. teilweise kraftfreie Magnetfelder. Zeitschrift Naturforschung Teil A 21, 1963 (1966)

A.I. Morozov, L.S. Solov'ev, Motion of charged particles in electromagnetic fields. Rev. Plasma Phys. **2**, 201 (1966)

F. Mottez, Exact nonlinear analytic Vlasov-Maxwell tangential equilibria with arbitrary density and temperature profiles. Phys. Plasmas **10**, 2501–2508 (2003)

F. Mottez, The pressure tensor in tangential equilibria. Annales Geophysicae **22**, 3033–3037 (2004)

L. Muschietti, I. Roth, C.W. Carlson, R.E. Ergun, Transverse instability of magnetized electron holes. Phys. Rev. Lett. **85**, 94–97 (2000)

H.E. Mynick, W.M. Sharp, A.N. Kaufman, Realistic Vlasov slab equilibria with magnetic shear. Phys. Fluids **22**, 1478–1484 (1979)

T. Neukirch, F. Wilson, M.G. Harrison, A detailed investigation of the properties of a Vlasov-Maxwell equilibrium for the force-free Harris sheet. Phys. Plasmas **16**(12), 122102 (2009)

C.S. Ng, A. Bhattacharjee, Bernstein-greene-kruskal modes in a three- dimensional plasma. Phys. Rev. Lett. 95.24, 245004 (2005)

C.S. Ng, A. Bhattacharjee, F. Skiff. Weakly collisional Landau damping and three-dimensional Bernstein-Greene-Kruskal modes: new results on old problems). Phys. Plasmas 13.5, 055903, 055903 (2006)

C.S. Ng, S.J. Soundararajan, E. Yasin, Electrostatic structures in space plasmas: stability of two-dimensional magnetic bernstein-greenekruskal modes. In: J. Heerikhuisen, G. Li, N. Pogorelov, G. Zank (eds.) American Institute of Physics Conference Series. vol. 1436. American Institute of Physics Conference Series, pp. 55–60 (2012)

R.B. Nicholson, Solution of the Vlasov equations for a Plasma in an externally uniform magnetic field. Phys. Fluids **6**, 1581–1586 (1963)

T.G. Northrop, The guiding center approximation to charged particle motion. Ann. Phys. **15**, 79–101 (1961)

T.G. Northrop, Adiabatic charged-particle motion. Rev. Geophys. Space Phys. **1**, 283–304 (1963)

A. Otto, K. Schindler, An energy principle for two-dimensional collisionless relativistic plasmas. Plasma Phys. Controll. Fusion **26**, 1525–1533 (1984)

E.V. Panov, A.V. Artemyev, R. Nakamura, W. Baumjohann, Two types of tangential magnetopause current sheets: Cluster observations and theory. J. Geophys. Res. (Space Physics) 116, A12204, A12204 (2011)

E.N. Parker, Spontaneous current sheets in magnetic fields: with applications to stellar x-rays. Spontaneous current sheets in magnetic fields: with applications to stellar x-rays. International Series in Astronomy and Astrophysics, vol. 1 (Oxford University Press, New York, 1994)

E.N. Parker, Sweet's mechanism for merging magnetic fields in conducting fluids. J. Geophys. Res. **62**(4), 509–520 (1957)

A.L. Peratt, advances in numerical modeling of astrophysical and space Plasmas. Astrophys. Space Sci. **242**, 93–163 (1996)
H.E. Petschek, Magnetic field annihilation. NASA Spec. Publ. **50**, 425 (1964)
T.D. Phan, G. Paschmann, Low-latitude dayside magnetopause and boundary layer for high magnetic shear 1. Structure and motion. J. Geophys. Res. **101**, 7801–7816 (1996)
D. Pines, D. Bohm, A Collective Description of electron interactions: II. Collective versus individual particle aspects of the interactions. Phys. Rev. **85**, 338–353 (1952)
J.H. Poynting, On the transfer of energy in the electromagnetic field. Philos. Trans. Royal Soc. Lond. **175**, 343–361 (1884)
E. Priest, T. Forbes, *Magnetic Reconnection* (Cambridge University Press, Cambridge, UK, 2000)
E. Priest. Magnetohydrodynamics of the Sun (2014)
P.L. Pritchett, Collisionless magnetic reconnection in an asymmetric current sheet. J. Geophys. Res. (Space Physics) 113, A06210, A06210 (2008)
K.B. Quest, F.V. Coroniti, Linear theory of tearing in a high-beta plasma. J. Geophys. Res. **86**, 3299–3305 (1981)
S.H. Rogers, E.C. Whipple, Generalized adiabatic theory applied to the magnetotail current sheet. Astrophys. Space Sci. **144**, 231–256 (1988)
M. Roth, J. de Keyser, M.M. Kuznetsova, Vlasov theory of the equilibrium structure of tangential discontinuities in space plasmas. Space Sci. Rev. **76**, 251–317 (1996)
R.Z. Sagdeev, Cooperative phenomena and shock waves in collisionless plasmas. Rev. Plasma Phys. **4**, 23 (1966)
H. Schamel, Electron holes, ion holes and double layers. Electrostatic phase space structures in theory and experiment. Phys. Reports **140**, 161–191 (1986)
H. Schamel, Hole equilibria in Vlasov-Poisson systems: A challenge to wave theories of ideal plasmas. Phys. Plasmas **7**, 4831–4844 (2000)
H. Schamel, Non-linear electrostatic plasma waves. J. Plasma Phys. **7**, 1–12 (1972a)
H. Schamel, Stationary solitary, snoidal and sinusoidal ion acoustic waves. Plasma Phys. **14**, 905–924 (1972b)
H. Schamel, Stationary solutions of the electrostatic Vlasov equation. Plasma Phys. **13**, 491–505 (1971)
H. Schamel, Theory of electron holes. Physica Scripta **20**, 336–342 (1979)
K. Schindler, *Physics of Space Plasma Activity*. (Cambridge University Press, 2007)
K. Schindler, J. Birn, Models of two-dimensional embedded thin current sheets from Vlasov theory. J. Geophys. Res. (Space Physics) **107**, 20–1 (2002)
J. Schmid-Burgk, Zweidimensionale selbstkonsistente Losungen der stationaren Wlassowgleichung fr Zweikomponentplasmen (1965). Max-Planck-Institut fr Physik und Astrophysik, Master's thesis
A. Sestero, Charge separation effects in the Ferraro-Rosenbluth cold plasma sheath model. Phys. Fluids **8**, 739–744 (1965)
A. Sestero, Self-consistent description of a warm stationary plasma in a uniformly sheared magnetic field. Phys. Fluids **10**, 193–197 (1967)
A. Sestero, Structure of plasma sheaths. Phys. Fluids **7**, 44–51 (1964)
A. Sestero, Vlasov equation study of plasma motion across magnetic fields. Phys. Fluids **9**, 2006–2013 (1966)
K. Shibata, S. Tanuma, Plasmoid-induced-reconnection and fractal reconnection. Earth Planets Space **53**(6), 473–482 (2001)
A. Spitkovsky, Particle acceleration in relativistic collisionless shocks: fermi process at last? Astrophys. J. **682**(L5), L5 (2008)
D.P. Stern, One-dimensional models of quasi-neutral parallel electric fields. Technical report (1981a)
D.P. Stern, one-dimensional models of quasi-neutral parallel electric fields. J. Geophys. Res. **86**, 5839–5860 (1981b)
A. Stockem, F. Fiuza, A. Bret, R.A. Fonseca, L.O. Silva, Exploring the nature of collisionless shocks under laboratory conditions. Sci. Reports **4**(3934), 3934 (2014)

S.-Y. Su, B.U.O. Sonnerup, On the equilibrium of the magnetopause current layer. J. Geophys. Res. (Space Physics) **76**, 5181–5188 (1971)

A. Suzuki, T. Shigeyama, A novel method to construct stationary solutions of the Vlasov-Maxwell system. Phys. Plasmas 15.4, 042107-+ (2008)

B. Svedung Wettervik, T.C. DuBois, T. Fulop, Vlasov modelling of laser-driven collisionless shock acceleration of protons. Phys. Plasmas 23.5, 053103, 053103 (2016)

P.A. Sweet, The neutral point theory of solar flares. In: B. Lehnert (ed.) Electromagnetic Phenomena in Cosmical Physics, vol. 6. (IAU Symposium, p. 123 1958)

M. Swisdak, B.N. Rogers, J.F. Drake, M.A. Shay, Diamagnetic suppression of component magnetic reconnection at the magnetopause. J. Geophys. Res. (Space Physics) 108, 1218, 1218 (2003)

E. Tassi, F. Pegoraro, G. Cicogna, Solutions and symmetries of forcefree magnetic fields. Phys. Plasmas 15.9, pp. 092113-+ (2008)

H. Tasso, G. Throumoulopoulos, Tokamak-like Vlasov equilibria. Eur. Phys. J. D **68**(175), 175 (2014)

J. Threlfall, T. Neukirch, C.E. Parnell, S. Eradat, Oskoui, Particle acceleration at a reconnecting magnetic separator. Astron. Astrophys. **574**(A7), A7 (2015)

D. Tong, Lectures on Classical Dynamics. http://www.damtp.cam.ac.uk/user/tong/dynamics.html (2004)

D. Tong, Lectures on Kinetic Theory. http://www.damtp.cam.ac.uk/user/tong/kinetic.html (2012)

L. Tonks, Trajectory-wise analysis of cylindrical and plane plasmas in a magnetic field and without collisions. Phys. Rev. 113, 400–407 (1959)

A. Vaivads, A. Retin?, J. Soucek, Yu. V. Khotyaintsev, F. Valentini, C.P. Escoubet, O. Alexandrova, M. Andr?, S.D. Bale, M. Balikhin et al., Turbulence heating ObserveR—satellite mission proposal. J. Plasma Phys. 82.5 (2016)

I.Y. Vasko, A.V. Artemyev, V.Y. Popov, H.V. Malova, Kinetic models of two-dimensional plane and axially symmetric current sheets: group theory approach. Phys. Plasmas 20.2, 022110, 022110 (2013)

I.Y. Vasko, O.V. Agapitov, F.S. Mozer, A.V. Artemyev, J.F. Drake, Electron holes in inhomogeneous magnetic field: Electron heating and electron hole evolution. Phys. Plasmas **23**(5), 052306 (2016)

G.E. Vekstein, N.A. Bobrova, S.V. Bulanov, On the motion of charged particles in a sheared force-free magnetic field. J. Plasma Phys. **67**, 215–221 (2002)

A.A. Vinogradov, I.Y. Vasko, A. V. Artemyev, E.V. Yushkov, A.A. Petrukovich, L.M. Zelenyi, Kinetic models of magnetic flux ropes observed in the Earth magnetosphere. Phys. Plasmas 23.7, 072901, 072901 (2016)

A.A. Vlasov, The vibrational properties of an electron gas. Phys. Uspekhi **10**(6), 721–733 (1968)

F. Wilson, T. Neukirch, A family of one-dimensional Vlasov-Maxwell equilibria for the force-free Harris sheet. Phys. Plasmas **18**(8), 082108 (2011)

M. Yamada, R. Kulsrud, H. Ji, Magnetic reconnection. Rev. Modern Phys. **82**, 603–664 (2010)

P.H. Yoon, A.T.Y. Lui, A class of exact two-dimensional kinetic current sheet equilibria. J. Geophys. Res. (Space Physics) 110, 1202-+ (2005)

P.H. Yoon, A.T.Y. Lui, R.B. Sheldon, On the current sheet model with distribution. Phys. Plasmas 13.10, 102108-+ (2006)

A.I. Zayed, Handbook of function and generalized function transformations. Math. Sci. Ref. Ser. (CRC Press, Boca Raton, FL, 1996), pp. xxiv+643

L.M. Zelenyi, H.V. Malova, A.V. Artemyev, V.Y. Popov, A. Petrukovich, Thin current sheets in collisionless plasma: equilibrium structure, plasma instabilities, and particle acceleration. Plasma Phys. Reports **37**(2), 118–160 (2011)

S. Zenitani, M. Hesse, A. Klimas, M. Kuznetsova, New measure of the dissipation region in Collisionless magnetic reconnection. Phys. Rev. Lett. 106.19, 195003, 195003 (2011)

Chapter 2
The Use of Hermite Polynomials for the Inverse Problem in One-Dimensional Vlasov-Maxwell Equilibria

Boltzmann's is still the most beautiful equation in the world, but Vlasov's isn't too shabby!

Cédric Villani

Much of the work in this chapter is drawn from Allanson et al. (2015, 2016, 2018).

2.1 Preamble

In this chapter, the aim is to make a contribution to the theory of exact equilibrium solutions to the Vlasov-Maxwell system, in 1D Cartesian geometry. In particular, we consider a solution method for the inverse problem in collisionless equilibria, namely that of calculating a VM equilibrium for a given macroscopic (fluid) equilibrium. Using Jeans' theorem (Jeans 1915), the equilibrium DFs are expressed as functions of the constants of motion, in the form of a stationary Maxwellian multiplied by an unknown function of the two conserved canonical momenta. In this case it is possible to reduce the inverse problem to inverting Weierstrass transforms, which we achieve by using expansions over Hermite polynomials. A sufficient condition on the pressure tensor is found which guarantees the convergence of the candidate solution when satisfied, and as a result the existence of velocity moments of all orders. This condition is obtained by elementary means, and it is clear how to put it into practice. We also argue that for a given pressure tensor for which our method applies, there always exists a non-negative DF for a sufficiently magnetised plasma. This argument is in fact proven for certain classes of DFs, and in the form of conjecture for others.

O. Allanson, *Theory of One-Dimensional Vlasov-Maxwell Equilibria*, Springer Theses, https://doi.org/10.1007/978-3-319-97541-2_2

2.2 Introduction

2.2.1 Hermite Polynomials in Fluid Closure

$$f = \frac{n(\boldsymbol{x}, t)}{(\sqrt{2\pi k_B T(\boldsymbol{x}, t)/m})^3} e^{-\boldsymbol{w}(\boldsymbol{x},t)^2/(2k_B T(\boldsymbol{x},t)/m)} \sum_{n=0}^{\infty} a^{(n)}(\boldsymbol{x}, t)\mathcal{H}^{(n)}(\boldsymbol{w}),$$

for $\mathcal{H}^{(n)}$ the n-dimensional Hermite "polynomial", and in fact a rank-n tensor, defined by

$$\mathcal{H}^{(n)}(\boldsymbol{w}) = \frac{(-)^n}{\mathcal{W}(\boldsymbol{w})} \frac{\partial^n}{\partial w_{i_1} \dots \partial w_{i_n}} \mathcal{W}(\boldsymbol{w}),$$
$$\text{s.t. } \mathcal{W}(\boldsymbol{w}) = \frac{1}{(2\pi)^{3/2}} e^{-w^2/(2k_B T(\boldsymbol{x},t)/m)}, \tag{2.1}$$

with each of the i_n-indices running over $\{x, y, z\}$. Note that—by the commutativity of partial derivatives—the labelling of the n-dimensional Hermite polynomials is somewhat degenerate, e.g. $H^{(2)}_{xy} = H^{(2)}_{yx} = w_x w_y$.

In this representation $a^{(n)}\mathcal{H}^{(n)}$ is the scalar product of two rank-n tensors, with the a 'coefficients' relating directly to the velocity moments of the DF, and as such they neatly 'index' the relationship between the particle distributions and certain macroscopic quantities:

$$\begin{aligned}
a^{(0)} &= 1 \iff \int f d^3v = n(\boldsymbol{x}, t),\\
\boldsymbol{a}^{(1)} &= (0, 0, 0) \iff \int w_i f d^3v = 0\\
a^{(2)}_{ij} &= \pi_{ij}/p, \iff \pi_{ij} = P_{ij}(\boldsymbol{x}, t) - \delta_{ij} p(\boldsymbol{x}, t)\\
a^{(3)}_{ijk} &= S_{ijk}/(p v_{\text{th}}) \iff \int w_i w_j w_k f d^3v = S_{ijk}(\boldsymbol{x}, t).\\
&\vdots
\end{aligned}$$

for δ_{ij} the Kronecker delta and S_{ijk} the heat flux tensor. By substituting this expanded form of the DF into Boltzmann's equation (Eq. (1.12)), multiplying by $\mathcal{H}^n(\boldsymbol{w})$ and then integrating over velocity space d^3v, Grad obtains an infinite hierarchy of differential equations that describe the spatial-temporal evolution of the $a^{(n)}$ coefficients, and in turn the moments of the DF. By truncating to third order (i.e. up to S_{ijk}), Grad then develops the "13-moment" equations for the variables n, $\boldsymbol{V}$, T, π_{ij} and $S_i = p v_{\text{th}} a^{(3)}_{ijj}$.

Grad uses Hermite polynomials (or generalisations thereof) in gas kinetic theory because of their orthogonality properties with respect to Gaussian functions, and this is what allows each term of order n in the expansion of the DF to be directly related to nth order velocity-space moments of the DF. It is for this very reason that Hermite polynomials have a long history in plasma physics.

2.2.2 Hermite Polynomials in VM Plasma Theory

The most typical approach in collisionless and weakly collisional plasma kinetic theory is to use expansions in 'scalar' Hermite polynomials, defined by

$$H_n(v) = (-1)^n e^{v^2} \frac{d^n}{dv^n} e^{-v^2}, \tag{2.2}$$

$$\int_{-\infty}^{\infty} H_m(v) H_n(v) e^{-v^2} dv = \delta_{mn} 2^n n! \sqrt{\pi}. \tag{2.3}$$

Hermite polynomials are a complete orthogonal set of polynomials for $f \in L^2(\mathbb{R}, e^{-v^2} dv)$ (Arfken and Weber 2001). That is to say that for any piecewise continuous f, such that

$$\int_{-\infty}^{\infty} |f|^2 e^{-v^2} dv < \infty, \tag{2.4}$$

then there exists an (infinite) expansion in Hermite polynomials, $\sum_{n=0}^{\infty} c_n H_n(v)$, such that

$$\lim_{k\to\infty} \int_{-\infty}^{\infty} \left| f - \sum_{n=0}^{k} c_n H_n(v) \right|^2 e^{-v^2} dv = 0. \tag{2.5}$$

where as Eqs. (2.2) and (2.4) are the standard definitions relevant to the use of Hermite polynomials, it will be of use in this work to consider the scaled function $H_n(v/(\sqrt{2}v_{\mathrm{th},s}))$, since Maxwellian DFs scale with $e^{-v^2/(2v_{\mathrm{th},s}^2)}$, as opposed to $e^{-v^2/(v_{\mathrm{th},s}^2)}$. This slight modification results in changes to Eqs. (2.2), (2.3), (2.4) and (2.5), easily achieved by substitution.

2.2.2.1 Hermite Polynomials in Velocity Space

As intimated above, expansions in Hermite polynomials are a natural choice for representing the velocity space structure of a DF in equilibrium and near-equilibrium plasmas, be the (near-)equilibrium collisional and hence (near-)thermal; or collisionless, and hence not necessarily (near-)thermal at all. Their suitability is epitomised by Eq. (2.3), and is demonstrated as follows.

First consider a quite general DF, written explicitly as a function over phase space $(\boldsymbol{x}, \boldsymbol{v}; t)$, and of the form

$$f_s(\boldsymbol{x}, \boldsymbol{v}, t) = \frac{n_s(\boldsymbol{x}, t)}{(\sqrt{2\pi} v_{\text{th},s}(\boldsymbol{x}, t))^3} e^{-v^2/(2(v_{\text{th},s}(\boldsymbol{x},t)^2))}$$
$$\times \sum_{ij} a_{ij}(\boldsymbol{x}, t) H_i\left(\frac{v_x}{\sqrt{2} v_{\text{th},s}(\boldsymbol{x}, t)}\right) H_j\left(\frac{v_y}{\sqrt{2} v_{\text{th},s}(\boldsymbol{x}, t)}\right), \quad (2.6)$$

where we define a time and space dependent thermal velocity by $v_{\text{th},s}(\boldsymbol{x}, t) = k_B T_s(\boldsymbol{x}, t)/m_s$. Expansions such as these are used in Hewett et al. (1976); Camporeale et al. (2006); Suzuki and Shigeyama (2008), for example. This form of the DF implies that a velocity space moment with respect to the (i, j)th-order Hermite polynomials is directly related to the (i, j)th-order coefficient of expansion,

$$\int f_s H_i\left(\frac{v_x}{\sqrt{2} v_{\text{th},s}}\right) H_j\left(\frac{v_y}{\sqrt{2} v_{\text{th},s}}\right) d^3 v \propto n_s(\boldsymbol{x}, t) a_{ijk}(\boldsymbol{x}, t).$$

A DF expanded in Hermite polynomials in the manner of Eq. (2.6) also possesses the feature that 'normal' velocity moments yield simple results, since the velocity space moments can be determined using the following definite integral (Gradshteyn and Ryzhik 2007),

$$\int_{-\infty}^{\infty} v^n e^{-v^2} H_n(v) dv = n! \sqrt{\pi}. \quad (2.7)$$

For example, the charge density and current density are directly related to the a_{ij} coefficients according to

$$\sigma(\boldsymbol{x}, t) \propto \sum_s q_s n_s a_{00},$$
$$j_x(\boldsymbol{x}, t) \propto \sum_s q_s n_s v_{\text{th},s} a_{10},$$
$$j_y(\boldsymbol{x}, t) \propto \sum_s q_s n_s v_{\text{th},s} a_{01}.$$

2.2.2.2 Hermite Polynomials in Momentum Space

The usefulness of Hermite polynomial expansions is not necessarily restricted to writing them as explicit functions of velocity space. If one considers VM equilibria, then as aforementioned the equilibrium DF is a function of phase space $(\boldsymbol{x}, \boldsymbol{v})$ *through* its dependence on the constants of motion. In such circumstances one could write the

DF as a stationary Maxwellian multiplied by an expansion in Hermite polynomials in the canonical momenta. For example, in the case of a 1D plasma such that $\nabla = (0, 0, \partial/\partial z)$, one could write

$$f_s = \frac{n_{0s}}{(\sqrt{2\pi} v_{\mathrm{th},s})^3} e^{-\beta_s H_s} \sum_{ij} a_{ij} H_i \left(\frac{p_{xs}}{\sqrt{2} m_s v_{\mathrm{th},s}} \right) H_j \left(\frac{p_{ys}}{\sqrt{2} m_s v_{\mathrm{th},s}} \right), \tag{2.8}$$

(e.g. see Abraham-Shrauner (1968), Channell (1976) for expansions such as these). Despite the fact that the Maxwellian factor, $e^{-\beta_s H_s}$, is a function of v^2, and the Hermite polynomials are functions of the momenta, one can still exploit the orthogonality properties of the Hermite polynomials. To see this, we can use the identity mentioned in Weisstein (2017)

$$H_j(x + y) = (H + 2x)^j, \quad \text{s.t.} \quad H^j := H_j(y). \tag{2.9}$$

The identity in Eq. (2.9), and proven below, is useful since we can associate $X = x + y$ with $p_{js} = m_s v_j + q_s A_j$. This allows us to re-write the DF from Eq. (2.8), and to separate the dependence on velocity and vector potential. Since the vector potential is a function of space (z) only, the phase-space variables have also been 'separated' allowing us to use results such as those explained in Sect. 2.2.2.1.

We now prove this identity, since it seems fairly non-standard, and the above reference cites personal communication as the source:

Proof We first make use of the generating function for Hermite polynomials (Arfken and Weber 2001)

$$\exp(2Xt - t^2) = \sum_{j=0}^{\infty} H_j(X) \frac{t^j}{j!}. \tag{2.10}$$

By substituting $X = x + y$ into Eq. (2.10) we see that

$$\begin{aligned} \exp(2(x + y)t - t^2) &= \sum_{j=0}^{\infty} H_j(x + y) \frac{t^j}{j!}, \\ &= \exp(2xt) \sum_{i=0}^{\infty} H_i(y) \frac{t^i}{i!}. \end{aligned}$$

Then, expanding $\exp(2xt)$ as an infinite series implies that

$$\sum_{i=0}^{\infty} \sum_{k=0}^{\infty} \frac{(2xt)^k}{k!} \frac{H_i(y) t^i}{i!} = \sum_{j=0}^{\infty} H_j(x + y) \frac{t^j}{j!}. \tag{2.11}$$

To isolate the $H_j(x+y)$ term, we now need to pick the terms such that $i+k=j$:

$$\frac{H_j(x+y)}{j!} = \sum_{k=0}^{j} \frac{(2x)^k}{k!(j-k)!} H_{j-k}(y), \tag{2.12}$$

$$\Longrightarrow H_j(x+y) = \sum_{k=0}^{j} \binom{j}{k} (2x)^k H_{j-k}(y), \tag{2.13}$$

$$\Longrightarrow H_j(x+y) = (H+2x)^j, \quad H^j := H_j(y). \tag{2.14}$$

2.2.3 Hermite Polynomials for Exact VM Equilibria

In the work by Abraham-Shrauner (1968), expansions in Hermite polynomials of the canonical momentum are used to solve the VM system for the case of 'stationary waves' in a manner like that to be described in this chapter. These correspond not to Vlasov equilibria, but rather to nonlinear waves that are stationary in the wave frame, as discussed in Sect. 1.3.6. Abraham-Shrauner considers a 1D plasma with only one component of current density, first in a general sense, and then considers three different magnetic field configurations. Alpers (1969) also presents a somewhat general discussion on the use of Hermite polynomials for 1D VM equilibria, and proceeds to consider models suitable for the magnetopause, with both one component of the current density, and with two. In the work by Channell (1976), two methods are presented for the solution of the inverse problem with neutral VM equilibria, by means of example. These two methods are inversion by Fourier transforms and—once again—expansion over Hermite polynomials respectively. Channell uses Hermite polynomials in the canonical momenta, but this time with two components of the current density, for the specific case of a magnetic field that is especially suitable to be considered as a stationary wave solution.

In contrast to Abraham-Shrauner (1968), Alpers (1969), Channell (1976), the works by Hewett (1976), Suzukis (2008) both consider the forwards problem in VM equilibria, and use Hermite polynomial expansions in velocity space, for 1D and 2D plasmas respectively. Hewett et al. (1976) assume a representation for the DF similar to that in Eq. (2.6) but with only one current density component, and ensure self-consistency with Maxwell's equations numerically, whereas Suzuki and Shigeyama (2008) use an analytical approach, e.g. demonstrating that the Hermite polynomial approach can reproduce known equilibria such as the Harris sheet (Harris 1962), and the Bennett Pinch (Bennett 1934).

To give a subset of (modern) examples outside the realm of equilibrium studies *per se*, Hermite polynomial expansions are used by Daughton (1999) to assess the linear stability of a Harris current sheet; by Camporeale et al. (2006) also on the linear stability problem, using a truncation method somewhat like that of (Grad, 1949), and managing to bypass the traditional approach of integrating over the 'unperturbed orbits' (Coppi et al. 1966; Drake and Lee 1977; Quest and Coroniti 1981; Daughton 1999); by Zocco (2015) on linear collisionless Landau damping (Landau 1946; Clement Mouhot and Cedric Villani 2011); and by Schekochihin et al. (2016) on the problem of the free-energy associated with velocity-space moments of the DF, in the problem of plasma turbulence.

2.2.3.1 Mathematical Criteria

Since a DF represents a probability (in phase space), it clearly must satisfy the property

$$f_s \geq 0 \, \forall \, \boldsymbol{x}, \boldsymbol{v}, t, \tag{2.15}$$

and since a DF found using a Hermite polynomial method could in principle include an infinite series of polynomials in momenta/velocity that does not represent a known function in closed form, it is by no means clear if Eq. (2.15) will be satisfied. This issue is recognised by Abraham-Shrauner (1968, Hewett (1976). Not only is the non-negativity in question, but it is not obvious whether a given expansion in Hermite polynomials even converges, and this question was also raised by Hewett et al. (1976). Finally, even if the Hermite expansion converges, it must-when multiplied by the Maxwellian factor—produce a DF for which velocity moments of all orders exist, as discussed in Sect. 1.2.1. In order to have full confidence in the Hermite polynomial method we need to address these issues of non-negativity, convergence, and the existence of moments.

Crucially, none of the above references tackle the questions of non-negativity and convergence of an infinite series of Hermite polynomials in a systematic way, or of the boundedness of the resultant DF. The method presented in this chapter should be seen as a rigorous extension, or generalisation, of the Hermite Polynomial discussed previously by these authors.

We should mention that the *reverse* questions are well established, i.e. if one *a priori* knows the DF in closed form, or at least if Eq. (2.4) is satisfied. In such circumstances, one can represent a given non-negative DF as a Maxwellian multiplied by an expansion in Hermite polynomials provided the g_s function grows at a rate below $e^{v^2/(4v_{\text{th},s}^2)}$ (Grad 1949; Widder 1951).

The structure of the rest of this chapter is as follows. Section 2.3 contains the details of a formal solution to the inverse problem, by using known methods of inverting *Weierstrass transforms* with possibly infinite series of Hermite polynomials. For the formal solution to meaningfully describe a DF however, these series must be convergent, positive and bounded. A sufficient condition for convergence that places a restriction on the pressure tensor is obtained in Sect. 2.4. In Sect. 2.5 we argue that for an appropriate pressure function, there always exists a positive DF, for a sufficiently magnetised plasma, including proofs for a certain class of function.

2.3 Formal Solution by Hermite Polynomials

It was demonstrated in Sect. 1.3.5.1 that the pressure tensor component P_{zz} can be seen as the 'key' to solving the inverse problem for VM equilibria. In a 1D $z-$dependent geometry, the inverse problem is encapsulated by Eq. (1.41), repeated below,

$$P_{zz}(A_x, A_y) = \frac{\beta_e + \beta_i}{\beta_e\beta_i}\frac{n_{0s}}{2\pi m_s^2 v_{\text{th},s}^2} \times \int_{-\infty}^{\infty}\int_{-\infty}^{\infty} e^{-\beta_s\left((p_{xs}-q_sA_x)^2+(p_{ys}-q_sA_y)^2\right)/(2m_s)} g_s(p_{xs}, p_{ys})dp_{xs}dp_{ys},$$

along with Ampère's Law and quasineutrality (in this chapter we shall assume strict neutrality),

$$\frac{\partial P_{zz}}{\partial A_x} = -\frac{1}{\mu_0}\frac{d^2A_x}{dz^2},$$
$$\frac{\partial P_{zz}}{\partial A_y} = -\frac{1}{\mu_0}\frac{d^2A_y}{dz^2},$$
$$\phi = 0.$$

The subsequent work in this chapter assumes that such a function, $P_{zz}(A_x, A_y)$, has been found. To make mathematical progress, we shall make the assumption that the P_{zz} function found is of either 'summative' or 'multiplicative' separability, i.e. that $P_{zz}(A_x, A_y)$ is of the form

$$P_{zz} = \frac{n_0(\beta_e + \beta_i)}{\beta_e\beta_i}\left(\tilde{P}_1(A_x) + \tilde{P}_2(A_y)\right) \text{ or } P_{zz} = \frac{n_0(\beta_e + \beta_i)}{\beta_e\beta_i}\tilde{P}_1(A_x)\tilde{P}_2(A_y). \tag{2.16}$$

The constants n_0, β_e and β_i are present in order to give the correct dimensions to the P_{zz} expression, in a species independent manner, such that the 'components' of the pressure, $\tilde{P}_1(A_x)$ and $\tilde{P}_2(A_y)$, are dimensionless. These assumptions are commensurate with

$$g_s = g_{1s}(p_{xs}; v_{\text{th},s}) + g_{2s}(p_{ys}; v_{\text{th},s}) \text{ or } g_s = g_{1s}(p_{xs}; v_{\text{th},s})g_{2s}(p_{ys}; v_{\text{th},s}), \tag{2.17}$$

respectively, and allow separation of variables according to

$$\tilde{P}_1(A_x) = \frac{1}{\sqrt{2\pi}m_s v_{\text{th},s}}\int_{-\infty}^{\infty} e^{-\beta_s(p_{xs}-q_sA_x)^2/(2m_s)} g_{1s}(p_{xs}; v_{\text{th},s})dp_{xs}, \tag{2.18}$$

$$\tilde{P}_2(A_y) = \frac{1}{\sqrt{2\pi}m_s v_{\text{th},s}}\int_{-\infty}^{\infty} e^{-\beta_s\left(p_{ys}-q_sA_y\right)^2/(2m_s)} g_{2s}(p_{ys}; v_{\text{th},s})dp_{ys}. \tag{2.19}$$

The separation constant is set to unity in the case of multiplicative separability, and zero in the case of additive separability, without loss of generality. We have included the parametric dependence on the thermal velocity, $v_{\mathrm{th},s}$, in the g_s functions to highlight the fact that the g_s functions must behave in such a way that the RHS of Eqs. (2.18) and (2.19) must, after integration, be independent of species as discussed in Sect. 1.3.5.4. This would be impossible if g_s did not depend on $v_{\mathrm{th},s}$.

The components of the pressure are now represented by 1D integral transforms of the unknown parts of the DF, namely Weierstrass transforms.

2.3.1 Weierstrass Transform

The Weierstrass transform, $u(x,t)$ of $u_0(y)$, is defined by

$$u(x,t) := \mathcal{W}\left[u_0\right](x,t) = \frac{1}{\sqrt{4\pi t}} \int_{-\infty}^{\infty} e^{-(x-y)^2/(4t)}\, u_0(y)\, dy, \tag{2.20}$$

see Bilodeau (1962) for example. This is also known as the Gauß transform, Gauß-Weiertrass transform and the Hille transform (Widder 1951). As the Green's function solution to the heat/diffusion equation,

$$\begin{aligned} \frac{\partial u}{\partial t} - \frac{\partial^2 u}{\partial x^2} &= 0, \\ \text{such that} \quad u(x,0) &= u_0(x), \ \forall x \in (-\infty,\infty), \\ \implies u(x,t) &= \mathcal{W}\left[u_0\right](x,t), \end{aligned}$$

$u(x,1)$ represents the temperature/density profile of an infinite rod one second after it was $u_0(x)$, see Widder (1951). Hence the Weierstrass transform of a positive function is itself a positive function.

2.3.2 Two Interpretations with Respect to Our Equations

Give or take some constant factors, Eqs. (2.18) and (2.19) express $\tilde{P}_1$ and $\tilde{P}_2$ as Weierstrass transforms of g_{1s} and g_{2s} respectively. To discuss this problem in generality, the following discussions in this chapter will make regular use of the subscript $j \in \{1,2\}$. This index will indicate the following components for the vector potential and canonical momenta,

$$\begin{aligned} (A_1, A_2) &:= (A_x, A_y), \\ (p_{1s}, p_{2s}) &:= (p_{xs}, p_{ys}) \end{aligned}$$

Otherwise, the indexing of P_1, P_2, g_{1s}, g_{2s} will remain "as is". As such the inverse problem is now characterised by the following equation,

$$\tilde{P}_j(A_j) = \frac{1}{\sqrt{2\pi}m_s v_{\text{th},s}} \int_{-\infty}^{\infty} e^{-\beta_s(p_{js}-q_sA_j)^2/(2m_s)} g_{js}(p_{js}; v_{\text{th},s}) dp_{js}$$

To be precise, there are two different interpretations of the equations that could be made here, namely:

Dimensionality retained and 'time' is a variable:

$$\tilde{P}_j(A_j) =: \mathcal{I}_j(\mathcal{A}_{js}) = \frac{1}{\sqrt{4\pi\varepsilon_s}} \int_{-\infty}^{\infty} e^{-(p_{js}-\mathcal{A}_{js})^2/(4\varepsilon_s)} g_{js}(p_{js}; \varepsilon_s) dp_{js}, \quad (2.21)$$

for $\varepsilon_s = m_s^2 v_{\text{th},s}^2/2$ and $\mathcal{A}_{js} = q_s A_j$. This first interpretation is depicted by Eq. (2.21) and casts the inverse problem in direct comparison with the Weierstrass transform, making a correspondence between space and time in the heat equation, (x, t), to $(\mathcal{A}_{js}, \varepsilon_s)$ in our inverse problem. However, one difference is that the g_s function must—at least parametrically—depend on 'time', ε_s, in contrast to the initial condition (i.e. time-independent function) that is part of the integrand in Eq. (2.20). We know that g_s must depend on a species-dependent parameter, i.e. ε_s, since the result of the integral (the LHS) must be independent of ε_s, in a similar vein to the discussion in Sect. 1.3.5.4.

Dimensionless variables and 'time' is fixed:

$$\tilde{P}_j\left(\text{sgn}(q_s)\delta_s A_j\right) =: \mathcal{J}_{js}(\tilde{A}_j; \delta_s) = \frac{1}{\sqrt{2\pi}} \int_{-\infty}^{\infty} e^{-(\tilde{p}_{js}-\tilde{A}_j)^2/2} \bar{g}_{js}(\tilde{p}_{js}; \delta_s) d\tilde{p}_{js}, \quad (2.22)$$

with $\text{sgn}(q_e) = -1$ and $\text{sgn}(q_i) = 1$, and for

$$\begin{aligned} \delta_s &= \frac{m_s v_{\text{th},s}}{eB_0L}, \\ \tilde{p}_{js} &= \frac{p_{js}}{m_s v_{\text{th},s}}, \\ \tilde{A}_j &= \frac{A_j}{B_0L} \\ \bar{g}_{js}(\tilde{p}_{js}; \delta_s) &= g_{js}(p_{js}; v_{\text{th},s}). \end{aligned}$$

The species-dependent magnetisation parameter, δ_s (e.g. see (Fitzpatrick, 2014)), is defined by

$$\delta_s = \frac{r_{Ls}}{L} = \frac{m_s v_{\text{th},s}}{eB_0L}.$$

It is the ratio of the thermal Larmor radius, $r_{Ls} = v_{\text{th},s}/|\Omega_s|$, to the characteristic length scale of the system, L. The gyrofrequency of particle species s is $\Omega_s = q_s B_0/m_s$. The magnetisation parameter is also known as the fundamental ordering parameter in gyrokinetic theory (see Howes (2006), Abel (2013) for example). In particle orbit theory, $\delta_s \ll 1$ implies that a guiding centre approximation will be applicable for that species, e.g. see Northrop (1961) and Sect. 1.1.1.

This second interpretation is depicted by Eq. (2.22) and once again casts the inverse problem in direct comparison with the Weiertrass transform, making a correspondence between space in the heat equation, x, to $\tilde{A}$ in our inverse problem. But in this case the 'time' is evaluated at $t = 1/2$. Since the LHS of Eq. (2.22) is now a function of δ_s, we have included the parametric dependence on δ_s in $\bar{g}_s$.

2.3.2.1 The 'Backwards Heat Equation'

The first interpretation is the one that I believe carries the most meaning for the problem considered in this thesis. Since the integral transform described by Eq. (1.41) must leave the LHS independent of species-dependent parameters, it makes sense that the transformed function, g_s, is not directly analogous to an initial condition. If g_s was an 'initial condition' and independent of 'time', ε_s, then the outcome of the evolution (transform) would surely give a time-dependent solution, i.e. one that depends on ε_s. But that is not what occurs. The correct analogy is to view the g_s function not as an initial condition, but *as the 'heat distribution'* ε_s *'seconds' ago, such that when evolved (transformed) forward by* ε_s *'seconds', the resultant 'heat distribution' is* P_{zz}. In that sense, we are considering the heat equation but with a *final condition*, as opposed to an initial condition: the 'backwards heat equation'. Similar topics are discussed in the 'backwards uniqueness of the heat equation' (see e.g. Lawrence 2010).

2.3.3 *Formal Inversion of the Weierstrass Transform*

Formally, the operator for the inverse Weierstrass transform is e^{-D^2}, with D the differential operator and the exponential suitably interpreted, see Eddington (1913); Widder (1954) for two different interpretations of this operator.

A second, and perhaps more computationally 'practical' method employs Hermite polynomials, see Bilodeau (1962). The Weierstrass transform of the nth Hermite polynomial $H_n(y/2)$ at $t = 1$ is x^n. Hence if one knows the coefficients of the Maclaurin expansion of $u(x, 1)$ in Eq. (2.20),

$$u(x, 1) = \sum_{j=0}^{\infty} \eta_j x^j,$$

then the Weierstrass transform can immediately be inverted to obtain the formal expansion

$$u_0(y) = \sum_{j=0}^{\infty} \eta_j H_j\left(y/2\right). \tag{2.23}$$

For this method to be useful in our problem, the pressure function must have a Maclaurin expansion that is convergent over all (A_x, A_y) space. Then, its coefficients of expansion must 'allow' the Hermite series to converge.

2.3.3.1 Formal Inversion of Our Problem

The following discussion applies to pressure functions of both summative and multiplicative form, with Maclaurin expansion representations (convergent over all (A_x, A_y) space) given by

$$\tilde{P}_1(A_x) = \sum_{m=0}^{\infty} a_m \left(\frac{A_x}{B_0 L}\right)^m, \quad \tilde{P}_2(A_y) = \sum_{n=0}^{\infty} b_n \left(\frac{A_y}{B_0 L}\right)^n, \tag{2.24}$$

with B_0 and L the characteristic magnetic field strength and spatial scale respectively. In line with the discussion on inversion of the Weierstrass transform in Sect. 2.3, we solve for g_s functions represented by the following expansions

$$g_{1s}(p_{xs}; v_{\mathrm{th},s}) = \sum_{m=0}^{\infty} C_{ms} H_m \left(\frac{p_{xs}}{\sqrt{2} m_s v_{\mathrm{th},s}}\right), \tag{2.25}$$

$$g_{2s}(p_{ys}; v_{\mathrm{th},s}) = \sum_{n=0}^{\infty} D_{ns} H_n \left(\frac{p_{ys}}{\sqrt{2} m_s v_{\mathrm{th},s}}\right), \tag{2.26}$$

with currently unknown species-dependent coefficients C_{ms} and D_{ns}. We cannot simply 'read off' the coefficients of expansion as in Eq. (2.23), since our integral equations are not quite in the 'perfect form' of Eq. (2.20). Upon computing the integrals of Eqs. (2.18) and (2.19) with the above expansions for g_s, we have

$$\tilde{P}_1(A_x) = \sum_{m=0}^{\infty} \left(\frac{\sqrt{2} q_s}{m_s v_{\mathrm{th},s}}\right)^m C_{ms} A_x^m, \quad \tilde{P}_2(A_y) = \sum_{n=0}^{\infty} \left(\frac{\sqrt{2} q_s}{m_s v_{\mathrm{th},s}}\right)^n D_{ns} A_y^n. \tag{2.27}$$

This result appears species dependent. However, to ensure self-consistency with quasineutrality ($n_i(A_x, A_y) = n_e(A_x, A_y)$)—as in Channell (1976, Harrison and Neukirch (2009), Wilson andNeukirch (2011)—we have to fix the pressure function to be species independent. It clearly must also match with the pressure function that maintains equilibrium with the prescribed magnetic field. The conditions to be derived here are critical for making a link between the macroscopic description of the equilibrium structure with the microscopic one of particles. These requirements imply—by the matching of Eqs. (2.24) and (2.27)—that

$$\left(\frac{\sqrt{2}q_s}{m_s v_{\mathrm{th},s}}\right)^m C_{ms} = \left(\frac{1}{B_0 L}\right)^m a_m \implies C_{ms} = \mathrm{sgn}(q_s)^m \left(\frac{\delta_s}{\sqrt{2}}\right)^m a_m, \quad (2.28)$$

$$\left(\frac{\sqrt{2}q_s}{m_s v_{\mathrm{th},s}}\right)^n D_{ns} = \left(\frac{1}{B_0 L}\right)^n b_n \implies D_{ns} = \mathrm{sgn}(q_s)^n \left(\frac{\delta_s}{\sqrt{2}}\right)^n b_n. \quad (2.29)$$

2.4 Mathematical Validity of the Method

2.4.1 Convergence of the Hermite Expansion

Here we find a sufficient condition that, when satisfied, guarantees that the Hermite series representations in (2.25) and (2.26) converge. This provides some answers to questions on the convergence of Hermite Polynomial representations of Vlasov equilibria dating back to Hewett et al. (1976), and implicit in the work of e.g. Alpers (1969), Channell (1976), Suzuki and Shigeyama (2008).

Theorem 1 *Consider a Maclaurin expansion of the form*

$$\tilde{P}_j(A_j) = \sum_{m=0}^{\infty} a_m \left(\frac{A_j}{B_0 L}\right)^m \quad (2.30)$$

that is convergent for all A_j. Then for $\varepsilon_s = m_s^2 v_{th,s}^2/2$ the function g_{js}, calculated in the inverse problem defined by the association

$$\tilde{P}_j(A_j) := \tilde{P}_{INT,j}(A_j) = \frac{1}{\sqrt{4\pi\varepsilon_s}} \int_{-\infty}^{\infty} e^{-(p_{js}-q_s A_j)^2/(4\varepsilon_s)} g_{js}(p_{js}; v_{th,s}) dp_{js}. \quad (2.31)$$

of the form

$$g_{js}(p_{js}; v_{th,s}) = \sum_{m=0}^{\infty} a_m \,\mathrm{sgn}(q_s)^m \left(\frac{\delta_s}{\sqrt{2}}\right)^m H_m\left(\frac{p_{js}}{\sqrt{2} m_s v_{th,s}}\right) \quad (2.32)$$

converges for all p_{js}, provided

$$\lim_{m\to\infty} \sqrt{m} \left|\frac{a_{m+1}}{a_m}\right| < 1/\delta_s, \quad (2.33)$$

in the case of a series composed of both even- and odd-order terms, or

$$\lim_{m\to\infty} m \left|\frac{a_{2m+2}}{a_{2m}}\right| < 1/(2\delta_s^2), \quad \lim_{m\to\infty} m \left|\frac{a_{2m+3}}{a_{2m+1}}\right| < 1/(2\delta_s^2), \tag{2.34}$$

in the case of a series composed only of even-, or odd-order terms, respectively.

Proof For a series composed of even- and odd-order terms, we have that

$$g_{js}(p_{js}; v_{\mathrm{th},s}) = \sum_{m=0}^{\infty} a_m \,\mathrm{sgn}(q_s)^m \left(\frac{\delta_s}{\sqrt{2}}\right)^m H_m\left(\frac{p_{js}}{\sqrt{2}m_s v_{\mathrm{th},s}}\right). \tag{2.35}$$

An upper bound on Hermite polynomials (see e.g. Sansone (1959)) is provided by the identity

$$|H_j(x)| < k\sqrt{j!}2^{j/2}\exp\left(x^2/2\right) \text{ s.t. } k = 1.086435\,. \tag{2.36}$$

This upper bound implies that

$$0 < |a_m|\left(\frac{\delta_s}{\sqrt{2}}\right)^m \left|H_m\left(\frac{p_s}{\sqrt{2}m_s v_{\mathrm{th},s}}\right)\right| < k|a_m|\delta_s^m\sqrt{m!}\exp\left(\frac{p_{js}^2}{4m_s^2 v_{\mathrm{th},s}^2}\right).$$

Let us now compose a series of the upper bounds,

$$g_{js,\mathrm{upper}} = k\exp\left(\frac{p_{js}^2}{4m_s^2 v_{\mathrm{th},s}^2}\right)\sum_{m=0}^{\infty}|a_m|\delta_s^m\sqrt{m!}.$$

By the use of the ratio test (Bartle and Sherbert 2000), a sufficient condition for convergence of $g_{js,\mathrm{upper}}$ is found by

$$\begin{aligned} &\lim_{m\to\infty}\left|\frac{a_{m+1}}{a_m}\right|\sqrt{m+1} < 1/\delta_s, \\ \Longrightarrow &\lim_{m\to\infty}\left|\frac{a_{m+1}}{a_m}\right|\sqrt{m} < 1/\delta_s, \end{aligned} \tag{2.37}$$

for a given $\delta_s \in (0, \infty)$. If the a_m satisfy the criteria in Eq. (2.37) then $g_{js,\mathrm{upper}}$ is a convergent series, and hence by the comparison test (Bartle and Sherbert 2000),

$$g_{js,\mathrm{absolute}} = \sum_{m=0}^{\infty}|a_m|\left(\frac{\delta_s}{\sqrt{2}}\right)^m\left|H_m\left(\frac{p_{js}}{\sqrt{2}m_s v_{\mathrm{th},s}}\right)\right|,$$

is a convergent series. This then implies that

$$\sum_{m=0}^{\infty} a_m \operatorname{sgn}(q_s)^m \left(\frac{\delta_s}{\sqrt{2}}\right)^m H_m\left(\frac{p_{js}}{\sqrt{2m_s}v_{\mathrm{th},s}}\right) (= g_{js}(p_{js}; v_{\mathrm{th},s}))$$

is an absolutely convergent series, and in turn a convergent series. We can now confirm that $g_{js}(p_{js}; v_{\mathrm{th},s})$ is a convergent series (Bartle and Sherbert 2000).

An analogous argument holds for those series with only even or odd order terms, with the ratio test giving

$$\lim_{m\to\infty}\left|\frac{a_{2m+2}}{a_{2m}}\right| m < 1/(2\delta_s^2), \quad \text{or} \quad \lim_{m\to\infty}\left|\frac{a_{2m+3}}{a_{2m+1}}\right| m < 1/(2\delta_s^2), \tag{2.38}$$

respectively. By the same argument as above, the comparison test implies that if the condition of (2.38) is satisfied, that since the series composed of upper bounds will converge, so must $g_{js}(p_{js})$. □

2.4.1.1 Decay Rate of the Coefficients

In order to get a better understanding of the meaning of Theorem 1, it is instructive to recapitulate the results in a continuous setting. One could imagine the modulus of the coefficients, $|a_m|$, as a subset of the codomain of a continuous function of the independent variable m,

$$\begin{aligned} &|a_m|,\ m = 0, 1, 2, \ldots \\ \to\ & a = a(m),\ m \in [0, \infty), \quad \text{s.t.}\ \ a(0) = |a_0|, a(1) = |a_1| \ldots . \end{aligned}$$

In this case, we require

$$a(m) = \mathcal{O}(a_{\mathrm{u}}(m)), \quad \text{s.t.}\ \ a_u(m) = (\delta_s^2 m)^{-m/2},$$

since the function a_{u} satisfies the restrictions of Eqs. (2.37) and (2.38), i.e.

$$\begin{aligned} \mathcal{O}\left(\left|\frac{a_{\mathrm{u}}(m+1)}{a_{\mathrm{u}}(m)}\right|\right) &= \frac{1}{\delta_s\sqrt{m}}, \\ \mathcal{O}\left(\left|\frac{a_{\mathrm{u}}(2m+2)}{a_{\mathrm{u}}(2m)}\right|\right) &= \frac{1}{2\delta_s^2 m}, \\ \mathcal{O}\left(\frac{a_{\mathrm{u}}(2m+3)}{a_{\mathrm{u}}(2m+1)}\right|\right) &= \frac{1}{2\delta_s^2 m}. \end{aligned}$$

Hence the modulus of the coefficients, $|a_m|$ must ‘fall below’ the graph of $(\delta_s^2 m)^{-m/2}$ for large m, and depicted in Fig. 2.1.

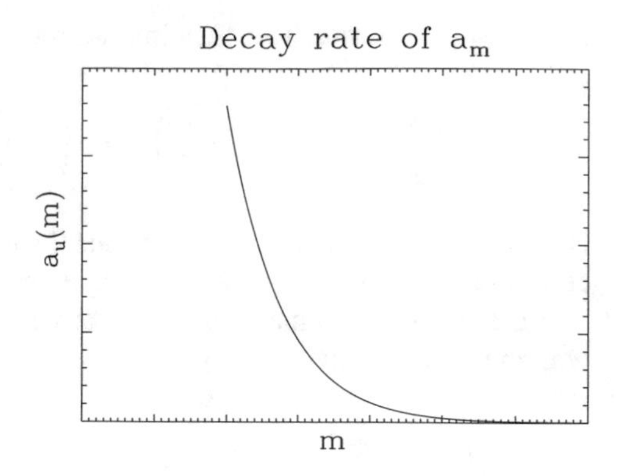

Fig. 2.1 Theorem 1 states that if the modulus of the coefficients, $|a_m|$, 'fall below' the graph of $(\delta_s^2 m)^{-m/2}$ as $m \to \infty$, then the Hermite series of Eq. (2.32) will converge

2.4.1.2 The Existence of Velocity Moments

Once the convergence of the Hermite polynomial is established, then one can begin to consider the boundedness of the DF, and the existence of velocity moments. If $g_s(p_s; v_{\text{th},s})$ is a convergent series, then by using Eq. (2.36) we see that

$$|g_{js}(p_{js}; v_{\text{th},s})| < \mathcal{L}_{js} \exp\left(\frac{p_{js}^2}{4m_s^2 v_{\text{th},s}^2}\right) \quad \forall\, p_{js},$$

and for $\mathcal{L}_{js}$ a finite, positive constant, independent of space and momentum. By now using the form of the DF from Eq. (1.37) and the separability conditions of Eq. (2.17), we see that

$$|f_s| < \exp\left[-(p_{xs} - q_s A_x)^2/(2m_s^2 v_{\text{th},s}^2) - (p_{ys} - q_s A_y)^2/(2m_s^2 v_{\text{th},s}^2) - v_z^2/(2v_{\text{th},s}^2)\right] \\ \times \left(\mathcal{L}_{xs} e^{p_{xs}^2/(4m_s^2 v_{\text{th},s}^2)} + \mathcal{L}_{ys} e^{p_{ys}^2/(4m_s^2 v_{\text{th},s}^2)}\right),$$

in the case of additive separability, or

$$|f_s| < \exp\left[-(p_{xs} - q_s A_x)^2/(2m_s^2 v_{\text{th},s}^2) - (p_{ys} - q_s A_y)^2/(2m_s^2 v_{\text{th},s}^2) - v_z^2/(2v_{\text{th},s}^2)\right] \\ \times \left(\mathcal{L}_{xs} \mathcal{L}_{ys} e^{p_{xs}^2/(4m_s^2 v_{\text{th},s}^2)} e^{p_{ys}^2/(4m_s^2 v_{\text{th},s}^2)}\right),$$

in the case of multiplicative separability. In either case, we see that boundedness in momentum space (and hence velocity space) is guaranteed. The reasoning is as follows. Since $p_{js} = m_s v_j + q_s A_j$, the arguments of the exponentials scale like

$$\exp\left(-\frac{v_j^2}{4v_{th,s}^2}\right), \tag{2.39}$$

in v_j velocity space. There is also a spatial dependence in the argument of the exponential, through $A_j(z)$, but this does not affect the velocity moment at a given z value. The scaling described by Expression (2.39) not only ensures boundedness, but guarantees that velocity moments of all order exist, since

$$\left| \int_{-\infty}^{\infty} v^k e^{-v^2/(4v_{th,s})^2} dv \right| < \infty \, \forall k \in 0, 1, 2, \ldots$$

2.4.1.3 Summary

In this Section we have shown that for a DF of the form

$$f_s(H_s, p_{xs}, p_{ys}) = \frac{n_{0s}}{(\sqrt{2\pi} v_{\text{th},s})^3} e^{-\beta_s H_s} g_s(p_{xs}, p_{ys}; v_{\text{th},s}),$$

with

$$g_s = g_{1s}(p_{xs}; v_{\text{th},s}) + g_{2s}(p_{ys}; v_{\text{th},s}) \text{ or } g_s = g_{1s}(p_{xs}; v_{\text{th},s}) g_{2s}(p_{ys}; v_{\text{th},s}),$$

and

$$g_{1s}(p_{xs}; v_{\text{th},s}) = \sum_{m=0}^{\infty} a_m \operatorname{sgn}(q_s)^m \left(\frac{\delta_s}{\sqrt{2}}\right)^m H_m \left(\frac{p_{xs}}{\sqrt{2} m_s v_{\text{th},s}}\right),$$

$$g_{2s}(p_{ys}; v_{\text{th},s}) = \sum_{n=0}^{\infty} b_n \operatorname{sgn}(q_s)^n \left(\frac{\delta_s}{\sqrt{2}}\right)^n H_n \left(\frac{p_{ys}}{\sqrt{2} m_s v_{\text{th},s}}\right),$$

the g_s functions are convergent provided the criteria on the growth rates of the coefficients of expansion from Theorem 1 are satisfied:

$$\lim_{m\to\infty} \sqrt{m} \left| \frac{a_{m+1}}{a_m} \right| < 1/\delta_s,$$

in the case of a series composed of both even- and odd-order terms, or

$$\lim_{m\to\infty} m \left| \frac{a_{2m+2}}{a_{2m}} \right| < 1/(2\delta_s^2), \quad \lim_{m\to\infty} m \left| \frac{a_{2m+3}}{a_{2m+1}} \right| < 1/(2\delta_s^2),$$

in the case of a series composed only of even-, or odd-order terms, respectively, and this in turn implies that velocity moments of the DF of all order exist.

2.5 Non-negativity of the Hermite Expansion

In this Section, we consider the non-negativity of the Hermite series representation of g_s—given by Eqs. (2.25) and (2.26)—and hence positivity of the DF. As such this Section responds to questions on the positivity of DF representation by Hermite polynomials raised by Abraham-Shrauner (1968), Hewett et al. (1976), and implicit in the work of e.g. Alpers (1969), Channell Suzuki and Shigeyama (1976, 2008).

2.5.1 *Possible Negativity of the Hermite Expansion*

For an example of a g_{js} function that is not necessarily always positive despite the pressure function being positive, consider a pressure function (e.g. from (Channell, 1976)) that is quadratic in the vector potential. In our notation, the pressure function considered by Channell is

$$\tilde{P} = \frac{1}{2}\left(a_0 + a_2\left(\frac{A_x}{B_0 L}\right)^2\right) + \frac{1}{2}\left(a_0 + a_2\left(\frac{A_y}{B_0 L}\right)^2\right),$$

for $a_0, a_2 > 0$. The resultant g_s function is of the form

$$g_s \propto \frac{1}{2}\left[a_0 + a_2\left(\frac{\delta_s}{\sqrt{2}}\right)^2 H_2\left(\frac{p_{xs}}{\sqrt{2}m_s v_{\text{th},s}}\right)\right] + \frac{1}{2}\left[a_0 + a_2\left(\frac{\delta_s}{\sqrt{2}}\right)^2 H_2\left(\frac{p_{ys}}{\sqrt{2}m_s v_{\text{th},s}}\right)\right].$$

Once these Hermite polynomials are expanded, and by substituting $p_{xs} = p_{ys} = 0$, we see that positivity of g_s is—for given values of a_0 and a_2—contingent on the size of δ_s,

$$g_s(0, 0) = a_0 - a_2\delta_s^2,$$
$$\therefore g_s(0, 0) \geq 0 \implies \delta_s^2 \leq \frac{a_0}{a_2}.$$

However, there is not necessarily anything 'special' about the origin, as compared to other points in momentum-space. For example, consideration of the pressure function

$$\tilde{P}_j = \left(a_0 + a_2\left(\frac{A_j}{B_0 L}\right)^2 + a_4\left(\frac{A_j}{B_0 L}\right)^4\right),$$

gives a g_{js} function that can, for given values of a_0, a_2, a_4 and for δ_s sufficiently large, be positive at $p_{js} = 0$, and negative at some other points.

It is worth considering how a g_{js} function that is negative for some p_{js} can transform in the manner of (2.18) and (2.19) to give a positive $\tilde{P}_j(A_j)$. One might expect that for certain values of A_j such that the Gaussian

$$e^{-(p_{js}-q_s A_j)^2/(4\varepsilon_s)}$$

is centred on the region in p_{js} space for which g_{js} is negative, that a negative value of $\tilde{P}_j(A_j)$ could be the result.

Essentially, the Gaussian will only 'successfully sample' a negative region of g_{js} to give a negative value of $\tilde{P}_j(A_j)$ if the Gaussian is narrow enough—for a given value of ε_s—to 'resolve' a negative patch of g_{js}. In other words, if the Gaussian is too broad, it won't 'see' the negative patches of g_{js}, and hence $\tilde{P}_j(A_j)$ will be positive. Hence the non-negativity of $\tilde{P}_j(A_j)$ is a restriction on the possible shape of g_{js}, and how that shape must scale with ε_s.

2.5.2 *Detailed Arguments*

When considering the non-negativity of the Hermite expansion, it is instructive to rewrite (2.31) in the form

$$\sum_{n=0}^{\infty} a_n \left(\mathrm{sgn}(q_s)\delta_s \tilde{A}_j\right)^n = \frac{1}{\sqrt{2\pi}} \int_{-\infty}^{\infty} e^{-(\tilde{p}_{js}-\tilde{A}_j)^2/2} \bar{g}_{js}(\tilde{p}_{js};\delta_s) d\tilde{p}_{js}, \tag{2.40}$$

by using the following associations

$$\tilde{A}_j = \frac{A_j}{B_0 L}, \quad \tilde{p}_{js} = \frac{p_s}{\sqrt{2\varepsilon_s}}, \quad g_{js}(p_{js};\varepsilon_s) = \bar{g}_{js}(\tilde{p}_{js};\delta_s).$$

The formal solution as an expansion in Hermite polynomials can be written as

$$\bar{g}_{js}(\tilde{p}_{js};\delta_s) = \sum_{n=0}^{\infty} a_n \mathrm{sgn}(q_s)^n \left(\frac{\delta_s}{\sqrt{2}}\right)^n H_n\left(\frac{\tilde{p}_{js}}{\sqrt{2}}\right). \tag{2.41}$$

We shall assume that the right-hand side of (2.41) represents a differentiable function. Note that the Gaussian in (2.40) is of fixed width $2\sqrt{2}$ (defined at $1/e$), in contrast to the Gaussian of variable width defined in (2.31).

2.5.2.1 Boundedness Below Zero of the Hermite Expansion

If the Hermite series satisfies the condition in Theorem 1 then it is convergent, so Eq. (2.36) gives

$$\left|\bar{g}_{js}(\tilde{p}_{js};\delta_s)\right| < \mathcal{L}_{js}e^{\tilde{p}_{js}^2/4}$$

for some finite and positive $\mathcal{L}_{js}$, determined by the sum of the (possibly infinite) series. Note that these bounds automatically imply integrability of f_s since as can be seen from Eq. (2.40), for some finite $L' > 0$, we have that $\left|\bar{g}_{js}(\tilde{p}_{js};\delta_s)\right| < L'e^{\tilde{p}_{js}^2/2}$ implies integrability, which is a less strict condition.

The bounds on $\bar{g}_{js}$ given above demonstrate that $\bar{g}_{js}$ can not tend to $\pm\infty$ for finite $\tilde{p}_{js}$. Hence, if it reaches $-\infty$ at all, it can only do so as $|\tilde{p}_{js}| \to \infty$. We argue however that the positivity of the pressure prevents the possibility of $\bar{g}_{js}$ being without a finite lower bound. The heuristic reasoning is as follows: the expression on the RHS of Eq. (2.40) treats—in the language of the heat/diffusion equation—the $\bar{g}_{js}$ function as the initial condition for a temperature/density distribution on an infinite 1-D line, and the left-hand side represents the distribution at some finite time later on. Were $\bar{g}_{js}$ to be unbounded from below, this would imply for our problem that a smooth 'temperature/density' distribution that is initially unbounded from below could, in some finite time, evolve into a distribution that has a positive and finite lower bound. This seems entirely unphysical since this would imply that an infinite negative 'sink' of heat/mass would somehow be 'filled in' above zero level in a finite time.

2.5.2.2 Proofs and Arguments by Contradiction

Here we give some technical remarks that support our claim that $\bar{g}_{js}$ (and hence g_{js}) is bounded below, using an argument by contradiction. First of all consider a smooth $\bar{g}_{js}$ function that is unbounded from below in positive momentum space. Then, depending on the number and nature of stationary points, either

- Case 1: There will be some $\tilde{p}_{j0,s}$ such that $\bar{g}_{js} < c < 0$ for all $\tilde{p}_{js} > \tilde{p}_{j0,s}$. This is a trivial statement if $\bar{g}_{js}$ has only a finite number of stationary points, whereas in the case of an infinite number of stationary points, all maxima of $\bar{g}_{js}$ for $\tilde{p}_{js} > \tilde{p}_{j0,s}$ must be 'away' from zero by a finite amount.
- Case 2: In this case the (infinite number of) maxima either can rise above zero, or tend to zero from below in a limiting fashion.

If $\bar{g}_{js}$ is of the type described in Case 1, then we can create an 'envelope' $g_{\text{env},j}$ for $\bar{g}_{js}$ such that $g_{\text{env},j} > \bar{g}_{js}$ for all $\tilde{p}_{js}$. The envelope we choose is

$$g_{\text{env},j} = \begin{cases} \mathcal{L}_{js}e^{\tilde{p}_{js}^2/4}, \text{ for } \tilde{p}_{js} \leq \tilde{p}_{j0,s}, \\ c \text{ for } \tilde{p}_{js} > \tilde{p}_{j0,s}. \end{cases} \tag{2.42}$$

The $\mathcal{L}_{js}e^{\tilde{p}_{js}^2/4}$ form for the profile is chosen because this represents the absolute upper bound for our convergent Hermite expansions, at a given $\tilde{p}_{js}$ as seen from Eq. (2.36). If we then substitute the $g_{\mathrm{env},j}$ function for $\bar{g}_{js}$ in Eq. (2.40) the integrals give combinations of error functions,

$$\frac{1}{\sqrt{2\pi}}\int_{-\infty}^{\infty} e^{-(\tilde{p}_{js}-\tilde{A}_j)^2/2}\bar{g}_{\mathrm{env},j}d\tilde{p}_{js} =$$
$$\frac{\mathcal{L}_{js}e^{\tilde{A}_j^2/2}}{\sqrt{2}}\left(\mathrm{erf}\left(\frac{\tilde{p}_{j0,s}-2\tilde{A}_j}{2}\right)+1\right)+\frac{c}{2}\left(\mathrm{erf}\left(\frac{\tilde{A}_j-\tilde{p}_{j0,s}}{\sqrt{2}}\right)+1\right)$$

from which it is seen that one obtains a negative result, i.e. c, as $\tilde{A}_j \to \infty$. This is a contradiction since the left-hand side of Eq. (2.40) is positive for all $\tilde{A}_j$. Hence we can discount the $\bar{g}_{js}$ functions of the variety described in Case 1, as we have a contradiction.

Case 2 is less simple to treat. The fact that there exists an infinite number of local minima and that the infimum of $\bar{g}_{js}$ is $-\infty$ implies that there exists an infinite sequence of points in momentum space, $\mathcal{S}_p = \{\tilde{p}_k : k = 1, 2, 3\ldots\}$, that are local minima of $\bar{g}_{js}$, such that $\bar{g}_{js}(\tilde{p}_{k+1}) < \bar{g}_{js}(\tilde{p}_k)$. Essentially there are an infinite number of minima 'lower than the previous one'. For sufficiently large $k = l$, we have that the magnitude of the minima is much greater than the width of the Gaussian, i.e.

$$|\bar{g}_{js}(\tilde{p}_l)| \gg 2\sqrt{2}.$$

In this case the only way that the sampling of $\bar{g}_{js}$ described by Eq. (2.40) could give a positive result for a Gaussian centred on the minima is if $\bar{g}_{js}$ rapidly grew to become sufficiently positive, in order to compensate the negative contribution from the minimum and its local vicinity. However, this seems to be at odds with the condition that $\bar{g}_{js}$ is smooth, since the function would have to rise in this manner for ever more negative values of the minima (and hence rise ever more quickly) as $k \to \infty$. We claim that this can not happen, and hence we discount the $\bar{g}_{js}$ functions of the variety described in Case 2.

Since there is no asymmetry in momentum-space in this problem, the arguments above hold just as well for for a $\bar{g}_{js}$ function that is unbounded from below in negative momentum space. It should be clear to see that if $\bar{g}_{js}$ can not be unbounded from below in either the positive or negative direction, then it can not be unbounded in both directions either.

2.5.2.3 Behaviour with Respect to the Magnetisation

If $\bar{g}_{js}$ (and hence g_{js}) is indeed bounded below then that means that one can always add a finite constant to g_{js} to make it positive, should the lower bound be known. However this constant contribution would directly correspond to raising the pressure (through the zeroth order Maclaurin coefficient a_0).

If we wish to consider a pressure function that is 'fixed', then we have a fixed a_0, and so it is not immediately obvious whether or not we can obtain a g_{js} that is positive over all momentum space. We have already seen some examples in Sect. 2.5.1 for which the sign of g_{js} depended on the value of δ_s.

Consider $\bar{g}_{js}$ evaluated at some particular value of $\tilde{p}_{js}$. We see from Eq. (2.41) that positivity requires

$$a_0 + c_1\delta_s + c_2\delta_s^2 + \cdots > 0,$$

for $c_1, c_2, \ldots$ finite constants. We also know that $a_0 > 0$ since $P_j(0) > 0$, i.e. the pressure is positive. This clearly demonstrates that positivity of g_{js} places some restriction on possible values of δ_s.

Let us now suppose that for a given value of δ_s, that there exists some regions in $\tilde{p}_{js}$ space where $\bar{g}_s < 0$. Our claim that $\bar{g}_{js}$ has a finite lower bound, combined with the expression in Eq. (2.41) implies that the $\bar{g}_s$ function is bounded below by a finite constant of the form $a_0 + \delta_s\mathcal{M}$, with

$$\mathcal{M} = \frac{1}{\sqrt{2}} \inf_{\tilde{p}_{js}} \sum_{n=1}^{\infty} a_n \mathrm{sgn}(q_s)^n \left(\frac{\delta_s}{\sqrt{2}}\right)^{n-1} H_n\left(\frac{\tilde{p}_{js}}{\sqrt{2}}\right),$$

and finite (and for inf the *infimum*, i.e. the greatest lower bound). By letting $\delta_s \to 0$ we see that $\bar{g}_{js}$ will converge uniformly to a_0, with

$$\lim_{\delta_s \to 0} \bar{g}_{js}(\tilde{p}_{js}, \delta_s) = a_0 > 0.$$

Hence, there must have existed some critical value of $\delta_s = \delta_c$ such that for all $\delta_s < \delta_c$ we have positivity of $\bar{g}_{js}$. Note that if the negative patches of $\bar{g}_{js}$ do not exist for any δ_s, then trivially $\delta_c = \infty$ as a special case.

2.5.3 Summary

To summarise, we claim—provided g_s is differentiable and convergent—that for values of the magnetisation parameter δ_s less than some critical value δ_c, according to $0 < \delta_s < \delta_c \leq \infty$, g_s is positive for any positive pressure function. The crucial step in this work was to prove/argue that g_s is bounded from below by a constant for all values of the momenta.

We have in fact proven this result for the class of g_s functions for which the number of stationary points is finite, or if infinite for which the stationary points are 'away' from zero by a finite amount. We have also presented arguments based on the differentiability of g_s, that support this result for other classes of g_s function.

2.6 Illustrative Case of the Use of the Method: Correspondence with the Fourier Transform Method

Here we give an example of the use of the solution method to a pressure function that was first discussed in Channell (1976). In that paper, Channell actually solved the inverse problem by the Fourier transform method, and showed that the solution was valid given certain restrictions on the parameters. We tackle the problem via the Hermite Polynomial method, and find that for the resultant DF to be convergent, we require exactly the same restrictions as Channell. This parity between the validity of the two methods is reassuring, and implies that the necessary restrictions on the parameters are in a sense 'method independent', and are the result of fundamental restrictions on the inversion of Weierstrass transformations.

The magnetic field considered by Channell can not be given analytically, but is of the form

$$\boldsymbol{B} = (B_x(z)\,,0\,,0), \quad \text{s.t.} \quad B_x(-\infty) = B_0, \tag{2.43}$$

and self-consistent with a pressure function

$$P_{zz} = P_0 e^{-\gamma \tilde{A}_y^2} \tag{2.44}$$

for P_0, B_0 and L characteristic values of the pressure, magnetic field and length scales, $\tilde{A}_y = A_y/(B_0 L)$ and $\gamma > 0$ dimensionless. The magnetic field and self-consistent number density profiles for this equilibrium are shown in Fig. 2.2, reproduced from Channell (1976). Note that the γ used by Channell has dimensions equivalent to $1/(B_0^2 L^2)$. We can now write the details of the inversion. The equation we must solve, for a DF given by

$$f_s = \frac{n_0}{(\sqrt{2\pi}v_{\mathrm{th},s})^3} e^{-\beta_s H_s} g_s(p_{ys};\, v_{\mathrm{th},s})$$

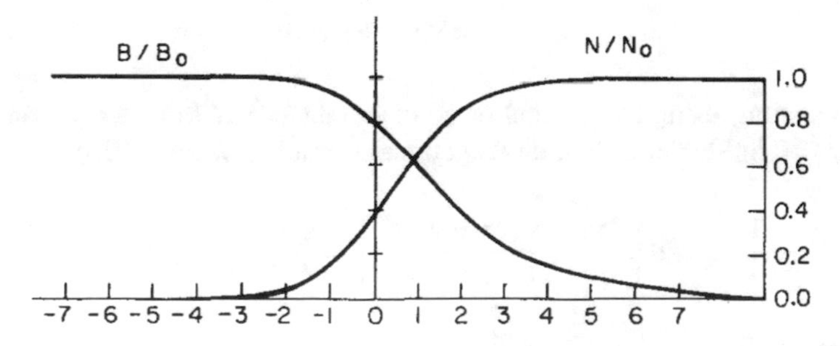

Fig. 2.2 A figure from Channell (1976) that displays the magnetic field and number density consistent with Eqs. (2.43) and (2.44). Image Copyright: AIP, *Physics of Fluids* **19**, 1541, (1976), copyright (1976), (reproduced with permission)

is

$$P_0 \exp\left(-\gamma \frac{A_y^2}{B_0^2 L^2}\right) = \frac{n_0(\beta_e + \beta_i)}{\beta_e \beta_i} \frac{1}{\sqrt{2\pi} m_s v_{\mathrm{th},s}} \int_{-\infty}^{\infty} \mathrm{e}^{-(p_{ys} - q_s A_y)^2/(2m_s^2 v_{\mathrm{th},s}^2)} g_s dp_{ys}.$$

We can immediately formally invert this equation as per the methods described in this Chapter, given the Maclaurin expansion of the pressure

$$P_{zz} = P_0 \sum_{m=0}^{\infty} a_{2m} \left(\frac{A_y}{B_0 L}\right)^{2m} \quad \text{s. t.} \quad a_{2m} = \frac{(-1)^m \gamma^m}{m!},$$

to give

$$g_s(p_{ys}) = \sum_{m=0}^{\infty} \left(\frac{\delta_s}{\sqrt{2}}\right)^{2m} a_{2m} H_{2m}\left(\frac{p_{ys}}{\sqrt{2} m_s v_{\mathrm{th},s}}\right).$$

Let us turn to the question of convergence. Theorem 1 states that if

$$\lim_{m\to\infty} m \left|\frac{a_{2m+2}}{a_{2m}}\right| < 1/(2\delta_s^2),$$

then the g_s function is convergent. This is readily seen to imply that γ must satisfy

$$\gamma < \frac{1}{2\delta_s^2},$$

for the Hermite series representation of g_s to be convergent. This condition is exactly equivalent to the one derived by Channell (Equation (28) in the paper). Note that now that we have established convergence for particular γ, then boundedness results follow as per the results in Sect. 2.4.1.2. One more question remains, namely how does the g_s function derived compare to the Gaussian $g_s(p_{ys})$ function derived by Channell

$$g_s \propto e^{-4\gamma^2 \delta_s^4 p_{ys}^2/(1-4\gamma^2\delta_s^4)}$$

(in our notation) using the method of Fourier transforms? In fact, one can see by setting $y = 0$ in Mehler's Hermite Polynomial formula (Watson 1933)

$$\frac{1}{\sqrt{1-\rho^2}} \exp\left[\frac{2xy\rho - (x^2+y^2)\rho^2}{1-\rho^2}\right] = \sum_{n=0}^{\infty} \frac{\rho^n}{2^n n!} H_n(x) H_n(y),$$

and using

$$H_m(0) = \begin{cases} 0 & \text{if } m \text{ is odd,} \\ (-1)^{m/2} m!/(m/2)! & \text{if } m \text{ is even,} \end{cases}$$

(see (Gradshteyn and Ryzhik, 2007) for example), that the Hermite series represents a Gaussian function in the range $|\rho| < 1$. This is equivalent to the condition derived above for convergence, $\gamma < 1/(2\delta_s^2)$. Hence, we have shown that for this specific example—solvable by using both Hermite polynomials and Fourier transforms—the two methods used to solve the inverse problem give equivalent functions with equivalent ranges of mathematical validity.

2.7 Summary

The primary result of this chapter is the rigorous generalisation of a solution method that exactly solves the 'inverse problem' in 1-D collisionless equilibria, for a certain class of equilibria. Specifically, given a pressure function, $P_{zz}(A_x, A_y)$, of a separable form, neutral equilibrium DFs can be calculated that reproduce the prescribed macroscopic equilibrium, provided P_{zz} satisfies certain conditions on the coefficients of its (convergent) Maclaurin expansion, and is itself positive.

The DF has the form of a Maxwellian modified by a function g_s, itself represented by—possibly infinite—series of Hermite polynomials in the canonical momenta. It is crucial that these series are convergent and positive for the solution to be meaningful. A sufficient condition was derived for convergence of the DF by elementary means, namely the ratio test, with the result a restriction on the rate of decay of the Maclaurin coefficients of P_{zz}. For DFs that are written as an expansion in Hermite polynomials, multiplied by a stationary Maxwellian, we have demonstrated that the necessary boundedness results follow.

We also argue that for such a pressure function that is also positive, that the Hermite series representation of the modification to the Maxwellian is positive, for sufficiently low values of the magnetisation parameter, i.e. lower than some critical value. This was actually proven for a certain class of g_s functions, and differentiability of g_s was assumed. It would be interesting in the future to investigate whether this critical value of the magnetisation parameter can be determined. It is also desirable that the result is proven for all reasonable function classes.

We have demonstrated the application of the solution method in Sect. 2.6. This particular example already has a known solution and range of validity in parameter space, obtained by a Fourier transform method in (Channell 1976). We obtain a solution with an alternate representation using the Hermite Polynomial method. The Hermite series obtained is shown to be equivalent to the representation obtained by Channell, and to have the exact same range of validity in parameter space. It is not clear if this equivalence between solutions obtained by the two different methods is true in general. Our problem is somewhat analagous to the heat/diffusion equation, and in that 'language' the question of the equivalence of solutions is related to the 'backwards uniqueness of the heat equation' (see e.g. (Evans 2010). The degree of similarity between our problem and the one described by Evans, and its implications, are left for future investigations.

Also, whilst we have assumed that the pressure is separable (either summatively or multiplicatively), the method should be adaptable in the 'obvious way' for pressures that are a 'superposition' of the two types. Interesting further work would be to see if the method can be adapted to work for pressure functions that are non-separable, i.e. of the form

$$P_{zz} = \sum_{m,n} \mathcal{C}_{mn} \left(\frac{A_x}{B_0 L} \right)^m \left(\frac{A_y}{B_0 L} \right)^n .$$

References

I.G. Abel, G.G. Plunk, E. Wang, M. Barnes, S.C. Cowley, W. Dorland, A.A. Schekochihin, Multiscale gyrokinetics for rotating tokamak plasmas: fluctuations, transport and energy flows. Reports Prog. Phys. **76**(11), 116201 (2013)

B. AbrahamspsShrauner, Exact, stationary wave solutions of the nonlinear vlasov equation. Phys. Fluids **11**, 1162–1167 (1968). June

O. Allanson, T. Neukirch, S. Troscheit, F. Wilson, From onedimensional fields to Vlasov equilibria: theory and application of Hermite polynomials. J. Plasma Phys. 82.3, p. 905820306 (2016)

O. Allanson, T. Neukirch, F.Wilson, S. Troscheit, An exact collisionless equilibrium for the Force-Free Harris Sheet with low plasma beta. Phys. Plasmas 22.10, 102116 (2015)

O. Allanson, S. Troscheit, T. Neukirch, On the inverse problem for Channell collisionless plasma equilibria. IMA J. Appl. Math., hxy026 (2018)

W. Alpers, Steady state charge neutral models of the magnetopause. Astrophys. Space Sci. **5**, 425–437 (1969). Dec

G.B. Arfken, H.J. Weber, *Mathematical Methods for Physicists*, 5th edn. (Harcourt/Academic Press, Burlington, MA, 2001)

R.G. Bartle, D.R. Sherbert, *Introduction to real Analysis* (JohnWiley & Sons Limited, Canada, 2000)

W.H. Bennett, Magnetically self-focussing streams. Phys. Rev. **45**(12), 890–897 (1934)

G.G. Bilodeau, TheWeierstrass transform and Hermite polynomials. Duke Math. J. **29**(2), 293–308 (1962)

E. Camporeale, G.L. Delzanno, G. Lapenta, W. Daughton, New approach for the study of linear Vlasov stability of inhomogeneous systems. Phys. Plasmas 13.9, 092110, p. 092110 (2006)

P.J. Channell, Exact Vlasov-Maxwell equilibria with sheared magnetic fields. Phys. Fluids **19**, 1541–1545 (1976). Oct

Clement Mouhot and Cedric Villani, On Landau damping. Acta Math. **207**(1), 29–201 (2011)

B. Coppi, G. Laval, R. Pellat, Dynamics of the geomagnetic Tail. Phys. Rev. Lett. **16**, 1207–1210 (1966). June

W. Daughton, The unstable eigenmodes of a neutral sheet. Phys. Plasmas **6**, 1329–1343 (1999). Apr

J.F. Drake, Y.C. Lee, Kinetic theory of tearing instabilities. Phys. Fluids **20**, 1341–1353 (1977). Aug

A.S. Eddington, On a formula for correcting statistics for the effects of a known error of observation. Mon. Not. Royal Astron. Soc. **73**, 359–360 (1913). Mar

L.C. Evans, Partial differential equations. Second. Vol. 19. Graduate Studies in Mathematics. Am. Math. Soc. Providence, RI, pp. xxii+749 (2010)

R. Fitzpatrick, *Plasma Physics: An Introduction* (CRC Press, Taylor & Francis Group, 2014)

H. Grad, On the kinetic theory of rarefied gases. Comm. Pure Appl. Math. **2**, 331–407 (1949)

I.S. Gradshteyn, I.M. Ryzhik, Table of integrals, series, and products. Seventh. (Elsevier/Academic Press, Amsterdam, pp. xlviii+1171, 2007)

E.G. Harris, On a plasma sheath separating regions of oppositely directed magnetic field. Nuovo Cimento **23**, 115 (1962)

M.G. Harrison, T. Neukirch, One-dimensional Vlasov-maxwell equilibrium for the force-free harris sheet. Phys. Rev. Lett. 102.13, pp. 135003-+ (2009)

D.W. Hewett, C.W. Nielson, D. Winske, Vlasov confinement equilibria in one dimension. Phys. Fluids **19**, 443–449 (1976). Mar

G.G. Howes, S.C. Cowley, W. Dorland, G.W. Hammett, E. Quataert, A.A. Schekochihin, Astrophysical gyrokinetics: basic equations and linear theory. Astrophys. J. **651**, 590–614 (2006). Nov

J.H. Jeans, On the theory of star-streaming and the structure of the universe. Mon. Not. Royal Astron. Soc. **76**, 70–84 (1915). Dec

L.D. Landau, On the vibrations of the electronic plasma. J. Phys. (USSR) **10**, 25–34 (1946)

T.G. Northrop, The guiding center approximation to charged particle motion. Ann. Phys. **15**, 79–101 (1961). July

K.B. Quest, F.V. Coroniti, Linear theory of tearing in a high-beta plasma. J. Geophys. Res. **86**, 3299–3305 (1981). May

G. Sansone, Orthogonal functions. Revised English ed. Translated from the Italian by A. H. Diamond; with a foreword by E. Hille. Pure and Applied Mathematics, Vol. IX. Interscience Publishers, Inc., New York; Interscience Publishers, Ltd., London, pp. xii+411 (1959)

A.A. Schekochihin, J.T. Parker, E.G. Highcock, P.J. Dellar, W. Dorland, G.W. Hammett, Phase mixing versus nonlinear advection in drift-kinetic plasma turbulence. J. Plasma Phys. **82**, 905820212, p. 47 (2016)

A. Suzuki, T. Shigeyama, A novel method to construct stationary solutions of the Vlasov-Maxwell system. Phys. Plasmas 15.4, p. 042107-+ (2008)

G.N. Watson, Notes on generating functions of polynomials: (2) Hermite Polynomials. J. London Math. Soc. s1-8.3, 194–199 (1933)

E.W. Weisstein, Hermite Polynomial. From MathWorld-A Wolfram Web Resource. http://mathworld.wolfram.com/HermitePolynomial.html (2017)

D.V. Widder, Necessary and sufficient conditions for the representation of a function by a Weierstrass transform. Trans. Am. Math. Soc. **71**, 430–439 (1951). Nov

D.V. Widder, The convolution transform. Bull. Am. Math. Soc. **60**(5), 444–456 (1954). Sept

F. Wilson, T. Neukirch, A family of one-dimensional Vlasov-Maxwell equilibria for the force-free Harris sheet. Phys. Plasmas **18**(8), 082108 (2011). Aug

A. Zocco, Linear collisionless Landau damping in Hilbert space. J. Plasma Phys. 81.4, 905810402, p. 049002 (2015)

Chapter 3
One-Dimensional Nonlinear Force-Free Current Sheets

We have to keep an eye on the electrons.

Thomas Neukirch

Much of the work in this chapter is drawn from Allanson et al. (2015, 2016).

3.1 Preamble

In this chapter we present new exact collisionless equilibria for a 1D nonlinear force-free magnetic field, namely the force-free Harris sheet. In contrast to previous solutions (Harrison and Neukirch 2009a; Wilson and Neukirch 2011; Abraham-Shrauner 2013; Kolotkov et al. 2015), the solutions that we present allow the plasma beta (β_{pl}) to take any value, and crucially values below unity for the first time. In the derivations of the equilibrium DFs it is found that the most typical approach of Fourier Transforms can not be applied, and so we use expansions in Hermite polynomials, making use of the techniques developed in Chap. 2. Using the convergence criteria developed therein, we verify that the Hermite expansion representation of the DFs are convergent for all parameter values. As shown in Chap. 2, this also implies boundedness, and the existence of velocity moments of all orders.

Despite the proven analytic convergence, initial difficulties in attaining numerical convergence mean that plots of the DF can be presented for the plasma beta only modestly below unity. In the effort to model equilibria with much lower values of the plasma, we use a new gauge for the vector potential, and calculate the DF consistent with this gauge, confirming the properties of convergence velocity moments. This new gauge makes attaining numerical convergence possible for lower values of the plasma beta, and we present results for $\beta_{pl} = 0.05$.

O. Allanson, *Theory of One-Dimensional Vlasov-Maxwell Equilibria*,
Springer Theses, https://doi.org/10.1007/978-3-319-97541-2_3

3.2 Introduction

Force-free equilibria, with fields defined by

$$\boldsymbol{j} \times \boldsymbol{B} = \frac{1}{\mu_0}(\nabla \times \boldsymbol{B}) \times \boldsymbol{B} = \boldsymbol{0}, \tag{3.1}$$

are of particular relevance to the solar corona (e.g. see (Priest and Forbes 2000; Wiegelmann and Sakurai 2012) and Fig. 3.1); current sheets in the Earth's magnetotail (e.g. (Vasko et al. 2014; Petrukovich et al. 2015)), the Earth's magnetopause (e.g. (Panov et al. 2011)) and in the Jovian magnetotail (e.g. (Artemyev et al. 2014)); other astrophysical plasmas (e.g. (Marsh 1996)); scrape-off layer currents in tokamaks (e.g. (Fitzpatrick 2007)), and 'Taylor-relaxed' magnetic fields in fusion experiments (e.g. (Taylor 1974, 1986)). Eq. (3.1) implies that the current density is everywhere-parallel to the magnetic field;

$$\mu_0 \boldsymbol{j} = \alpha(\boldsymbol{x})\boldsymbol{B}, \tag{3.2}$$

or zero in the case of potential fields, and with α the *force-free parameter*. If $\nabla\alpha \neq 0$ then the force-free field is nonlinear, whereas a constant α corresponds to a linear force-free field. Note that

$$\nabla \cdot (\nabla \times \boldsymbol{B}) = 0 \implies \boldsymbol{B} \cdot \nabla\alpha = 0,$$

and hence α is a constant along a magnetic field line, but will vary from field line to field line in the case of nonlinear force-free fields. Extensive discussions of force-free fields are given in Sakurai (1989) and Marsh (1996).

3.2.1 *Force-Free Equilibria and the Plasma Beta*

Equation (3.1) presents the force-free condition in purely geometric terms, i.e. an equilibrium force-free magnetic field has field lines obeying certain geometrical constraints, such that a particular combination of spatial derivatives vanish. In order to gain some physical insight, consider a generic plasma equilibrium (in the absence of a gravitational potential),

$$\nabla \cdot \boldsymbol{P} = \sigma \boldsymbol{E} + \boldsymbol{j} \times \boldsymbol{B}. \tag{3.3}$$

Next, normalise each of the quantities according to

$$\nabla \cdot \boldsymbol{P} = \frac{p_0}{L_P} \tilde{\nabla} \cdot \tilde{\boldsymbol{P}},$$
$$\sigma \boldsymbol{E} = \sigma_0 E_0 \tilde{\sigma} \tilde{\boldsymbol{E}},$$
$$\boldsymbol{j} \times \boldsymbol{B} = \frac{B_0^2}{\mu_0 L_B} \tilde{\boldsymbol{j}} \times \tilde{\boldsymbol{B}},$$

for L_P, L_B typical values of the length scales associated with the pressure and magnetic fields respectively; and with p_0, σ_0, E_0, B_0 typical values of the thermal pressure, charge density, electric and magnetic field respectively. Furthermore, since $\boldsymbol{E} = -\nabla\phi$ and $\nabla^2\phi = -\sigma/\epsilon_0$, we define

$$\sigma_0 = -\frac{\epsilon_0 \phi_0}{L_\phi^2},$$
$$E_0 = -\frac{\phi_0}{L_\phi},$$
$$\text{s.t.} \quad \phi_0 = \frac{k_B T_0}{e},$$

for T_0 a typical value of the temperature, and L_ϕ the length scale associated with the scalar potential. Written in dimensionless form, the force balance equation (Eq. (3.3)) can now be written as

$$\frac{\beta_{pl}}{2} L_B \left[\frac{1}{L_P} \tilde{\nabla} \cdot \tilde{\boldsymbol{P}} - \frac{1}{L_\phi} \frac{\lambda_D^2}{L_\phi^2} \tilde{\sigma} \tilde{\boldsymbol{E}} \right] = \tilde{\boldsymbol{j}} \times \tilde{\boldsymbol{B}},$$

for $\beta_{pl} = 2\mu_0 p_0 / B_0^2$ the plasma beta, and $\lambda_D = \sqrt{\epsilon_0 k_B T_0 / (n_0 e^2)}$ the Debye radius. Note that we have made use of $p_0 = n_0 k_B T_0$. This equation demonstrates that—in principle -

$$\beta_{pl} \ll 1 \not\Longleftrightarrow \boldsymbol{j} \times \boldsymbol{B} = \boldsymbol{0},$$

for $\not\Longleftrightarrow$ to read as 'not equivalent', i.e. force free equilibria need not necessarily have a vanishing plasma beta, or vice versa. However, we see that for a quasineutral plasma in which $\epsilon = \lambda_D / L_\phi \ll 1$, the second term on the LHS is—for a given value of β_{pl}—almost certainly of a lower order than the first term on the LHS, due to the ϵ^2 dependence. Hence we see that for a quasineutral equilibrium

$$\frac{\beta_{pl}}{2} \frac{L_B}{L_P} \tilde{\nabla} \cdot \tilde{\boldsymbol{P}} = \tilde{\boldsymbol{j}} \times \tilde{\boldsymbol{B}},$$

and so it would now seem fair to say that $\beta_{pl} \ll 1 \iff \boldsymbol{j} \times \boldsymbol{B} = \boldsymbol{0}$ for a quasineutral equilibrium, unless the thermal pressure varies with respect to very fine length scales. For a similar discussion to the above, including the gravitational acceleration but not the electric field, see Neukirch (2005). Figure 3.1 is reproduced from Gary (2001)

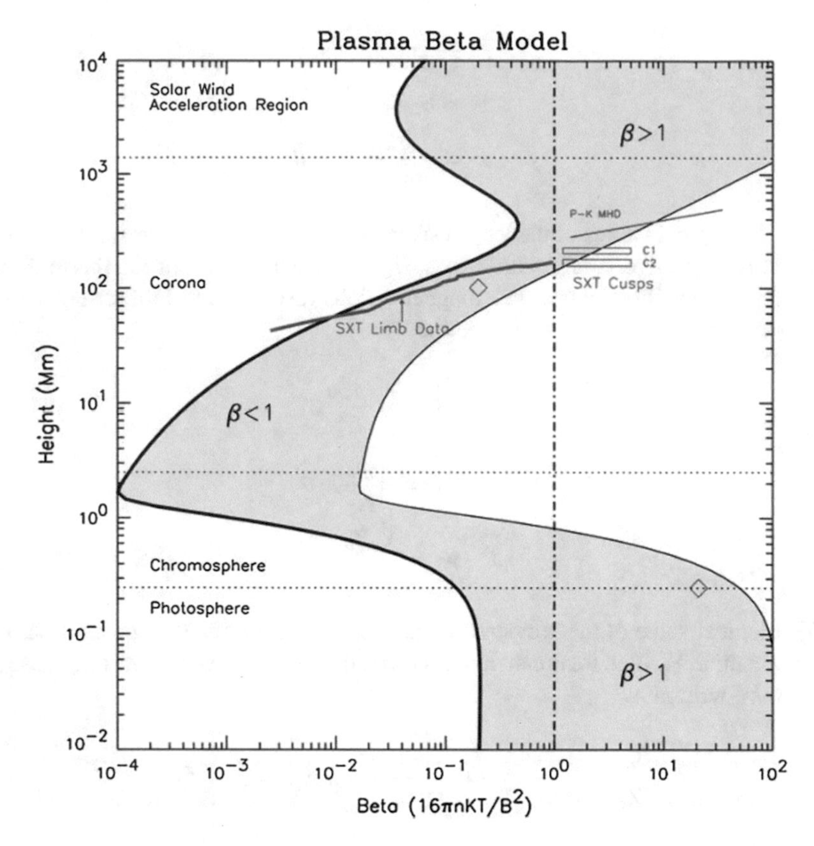

Fig. 3.1 A figure from Gary (2001) that displays a representative β_{pl} model over solar active regions, derived from a range of sources. Image Copyright: Springer, *Solar Physics* **203**, 1, (October 2001), pp. 71–86., copyright (2001), (reproduced with permission)

and shows a model for the plasma beta in the solar atmosphere, compiled from observational data. The figure demonstrates that β_{pl} can take sub-unity and vanishing values in the solar chromosphere and the corona, as well as values above one (contrary to the most typical assumptions). As such, much of the solar corona magnetic field is modelled as force-free (Wiegelmann and Sakurai 2012).

3.2.2 1D Force-Free Equilibria

1D magnetic fields can be represented without loss of generality by

$$\boldsymbol{B} = \big(B_x(z), B_y(z), 0\big) = \left(-\frac{dA_y}{dz}, \frac{dA_x}{dz}, 0\right). \tag{3.4}$$

The force-free condition then implies that

$$\boldsymbol{j} \times \boldsymbol{B} = \boldsymbol{0} \implies \frac{d}{dz}\left(\frac{B_x^2}{2\mu_0} + \frac{B_y^2}{2\mu_0}\right) = 0, \tag{3.5}$$

and hence the magnetic field is necessarily of uniform magnitude. Considering the equation of motion for a quasineutral plasma, now given by

$$\frac{d}{dz}\left(P_{zz} + \frac{B^2}{2\mu_0}\right) = 0,$$

we see that the thermal pressure is also of constant magnitude,

$$\frac{d}{dz}P_{zz} = 0 \implies P_{zz} = \text{const.} \tag{3.6}$$

As demonstrated in Sect. 1.3.5.1, the (assumed) existence of a VM equilibrium implies—through the dependence of the DF on the constants of motion—that the pressure tensor is a function of the vector and scalar potentials. Hence, we see that for a quasineutral plasma in which $\phi_{qn} = \phi(A_x, A_y)$, the force-free equilibrium fields correspond to a *trajectory*, $\boldsymbol{A}_{ff}(z) = (A_x(z), A_y(z), \phi_{qn}(A_x(z), A_y(z)))$, that is itself a *contour*;

$$\frac{d}{dz}P_{zz}(A_x(z), A_y(z)) = 0, \tag{3.7}$$

of the *potential*, P_{zz} (Harrison and Neukirch 2009a, b). As such, the construction of a P_{zz} function that satisfies Eq. (3.7), given some $(A_x(z), A_y(z)$ is the first step in the inverse method for 1D force-free equilibria.

In fact, Eq. (3.7) compactly defines the entire macroscopic problem, since

$$\frac{\partial P_{zz}}{\partial \boldsymbol{A}} = \boldsymbol{j}, \tag{3.8}$$

implies that

$$\begin{aligned} \frac{d}{dz}P_{zz}(A_x(z), A_y(z)) &= \underbrace{\frac{\partial P_{zz}}{\partial A_x}}_{j_x}\underbrace{\frac{dA_x}{dz}}_{B_y} + \underbrace{\frac{\partial P_{zz}}{\partial A_y}}_{j_y}\underbrace{\frac{dA_y}{dz}}_{-B_x} = 0, \\ &= j_x B_y - j_y B_x, \\ &= (\boldsymbol{j} \times \boldsymbol{B})_z. \end{aligned} \tag{3.9}$$

This demonstrates that—in a 1D quasineutral plasma—the existence of a VM equilibrium that is self-consistent with a spatially uniform pressure tensor directly implies that the magnetic field is force-free.

3.2.2.1 Pressure Tensor Transformation Theory

The inverse problem is not only non-unique regarding the form of the DF for a particular macroscopic equilibrium (as discussed in Sect. 1.3.5), but also for the form of $P_{zz}(A_x, A_y)$ for a particular magnetic field. Given a specific force-free magnetic field, i.e. a specific $\left(A_x, A_y\right)$, and a known P_{zz} that satisfies Eqs. (3.7) and (3.8), one can construct infinitely many new $\bar{P}_{zz}$ functions that also satisfy them;

$$\bar{P}_{zz} = \frac{1}{\psi'(P_{ff})}\psi(P_{zz}), \tag{3.10}$$

for differentiable and non-constant ψ, provided the LHS is positive, and for which the value of P_{zz} evaluated on the force-free contour, $\boldsymbol{A}_{ff}$, is the constant, P_{ff} (Harrison and Neukirch 2009b). These $\bar{P}_{zz}$ functions maintain a force-free equilibrium with the *same magnetic field* as P_{zz}, since

$$\left.\frac{\partial \bar{P}_{zz}}{\partial \boldsymbol{A}}\right|_{\boldsymbol{A}_{ff}} = \left.\frac{1}{\psi'(P_{ff})}\frac{\partial \psi}{\partial P_{zz}}\frac{\partial P_{zz}}{\partial \boldsymbol{A}}\right|_{\boldsymbol{A}_{ff}} = \left.\frac{\partial P_{zz}}{\partial \boldsymbol{A}}\right|_{\boldsymbol{A}_{ff}} = \boldsymbol{j}_{ff},$$

for $\boldsymbol{j}_{ff}$ the current density derived from $\boldsymbol{A}_{ff}$.

3.3 Force-Free Current Sheet VM Equilibria

Since current sheets are extremely important for reconnection studies (e.g. see (Priest and Forbes 2000)), and it is appropriate in many circumstances to model the magnetic field as force-free, a natural step is to construct VM equilibria for force-free current sheets. The archetypal 1D current sheet structure used to model reconnection is the Harris sheet (Harris 1962) (see Sect. 1.3.2.1),

$$\boldsymbol{B} = B_0(\tanh(z/L), 0, 0),$$

for which an exact VM equilibrium DF is well-known. However, the Harris sheet has $\boldsymbol{j} \perp \boldsymbol{B}$ and hence is not force-free, with thermal pressure gradients balancing those of the magnetic pressure. It is possible to approximate a force-free field with the addition of a uniform guide field

$$\boldsymbol{B} = (B_{x0}\tanh(z/L), B_{y0}, 0),$$

for B_{x0}, B_{y0} constants. This magnetic field configuration is frequently chosen as the initial condition in PIC simulations of magnetic reconnection (e.g. see (Pritchett and Coroniti 2004)), and the VM equilibrium is easily implemented since it is the same as that for the Harris sheet (Eq. (1.25)).

In principle, this magnetic field does approach a force free configuration for $B_{y0} \gg B_{x0}$, since $\boldsymbol{j}$ is approximately parallel to $\boldsymbol{B}$. However, the current density, j_y, is completely independent of the magnitude of the guide field, and so it is quite unlike an exact force-free field, for which the field-aligned current is related to the shear of the magnetic field. The equilibrium force balance is still maintained by the balance between gradients in the thermal pressure and the magnetic pressure,

$$\frac{1}{2\mu_0}\left(B_{x0}^2\tanh^2\left(\frac{z}{L}\right)+B_{y0}^2\right),$$

unlike for an exact force-free field. Finally, the addition of the guide field adds no extra free energy to the system (Harrison 2009). Hence it is of value to consider VM equilibria self-consistent with exact force-free magnetic fields because of their distinct physical nature, with one motivation in mind to see how these differences affect the magnetic reconnection process.

As discussed in e.g. Bobrova et al. (2001), Vekstein et al. (2002), Eq. (3.5) implies that a 1D force-free field can be written without loss of generality as

$$\boldsymbol{B}(z) = B_0(\cos(S(z)), \sin(S(z)), 0), \tag{3.11}$$

where $S(z) = \int \alpha(z)dz$, for α defined in Eq. (3.2). 1D linear force-free fields then, necessarily, have $S(z)$ as a linear function of z, i.e. $S_0z + S_1$. As a result, Eq. (3.11) then implies that that the magnetic field configuration for linear force-free fields will be periodic in the z direction, and hence there will be an infinite sequence of current sheet structures,

$$\boldsymbol{j} = \frac{-B_0S_0}{\mu_0 L}(\sin(S_0z + S_1), \cos(S_0z + S_1), 0).$$

Figure 3.2 displays the magnetic field from Eq. (3.11), and its current density, for $S(z) = z - \pi/2$.

In contrast to linear force-free fields, nonlinear force-free fields admit—in principle—all reasonable varieties of differentiable $S(z)$ functions, and hence are able to describe single, localised and intense current sheet structures.

3.3.1 Known VM Equilibria for Force-Free Magnetic Fields

The first VM equilibria self-consistent with linear force-free fields were found approximately fifty years ago, (Moratz and Richter 1966; Sestero 1967), with further examples of equilibria in Channell (1976), Bobrova and Syrovatskiĭ (1979), Correa-Restrepo and Pfirsch (1993), Attico and Pegoraro (1999), Bobrova et al. (2001) (note that Channell (1976); Attico and Pegoraro (1999) don't actually make the connection to force-free fields, but write down DFs that are self-consistent with such fields).

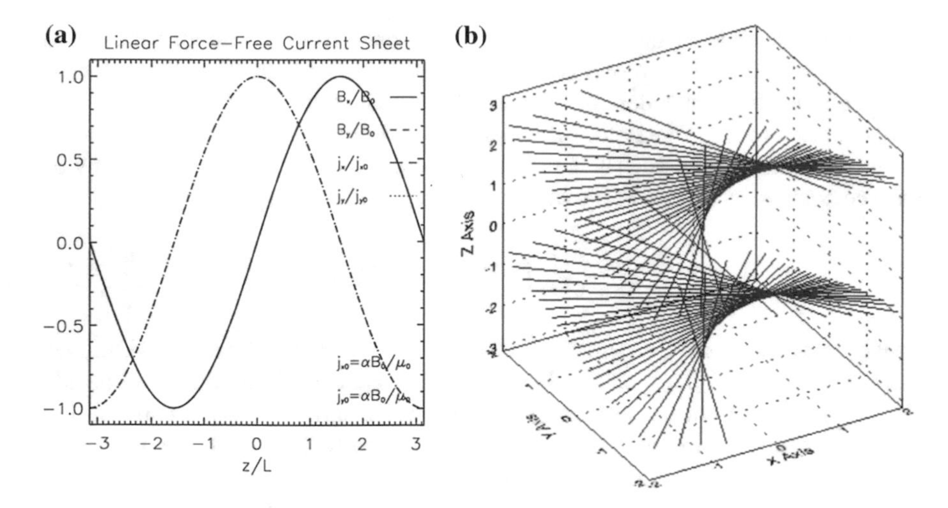

Fig. 3.2 **a** shows the magnetic field and current density components for a linear force-free field with $S(z) = z - \pi/2$. **b** shows the magnetic field lines. Images copyright: M.G. Harrison's PhD thesis (Harrison 2009), (reproduced with permission)

A limited number of PIC studies with exact VM equilibria for linear force-free fields as initial conditions have been conducted in Bobrova et al. (2001), Li et al. (2003), Nishimura et al. (2003), Sakai and Matsuo (2004), Bowers and Li (2007), Harrison (2009).

In contrast, exact VM equilibria for nonlinear force-free fields were only discovered in Harrison and Neukirch (2009a) (see also Neukirch et al. 2009), with subsequent solutions in Wilson and Neukirch (2011), Abraham-Shrauner (2013), Kolotkov et al. (2015), and 'nearly force-free' equilibria in Artemyev (2011). As a result, the investigations of the linear and nonlinear dynamics of such configurations are at an early stage (Harrison 2009; Wilson 2013; Wilson et al. 2017), with the first fully kinetic simulations of collisionless reconnection with an initial condition that is an exact Vlasov solution for a nonlinear force-free field conducted by (Wilson et al. 2016), and using the DF derived by Harrison and Neukirch (2009a).

3.3.1.1 The Force-Free Harris Sheet

The nonlinear force-free VM equilibrium solutions derived by Harrison and Neukirch (2009a), Wilson and Neukirch (2011), Kolotkov et al. (2015) are self-consistent with the force-free Harris sheet (FFHS), defined by

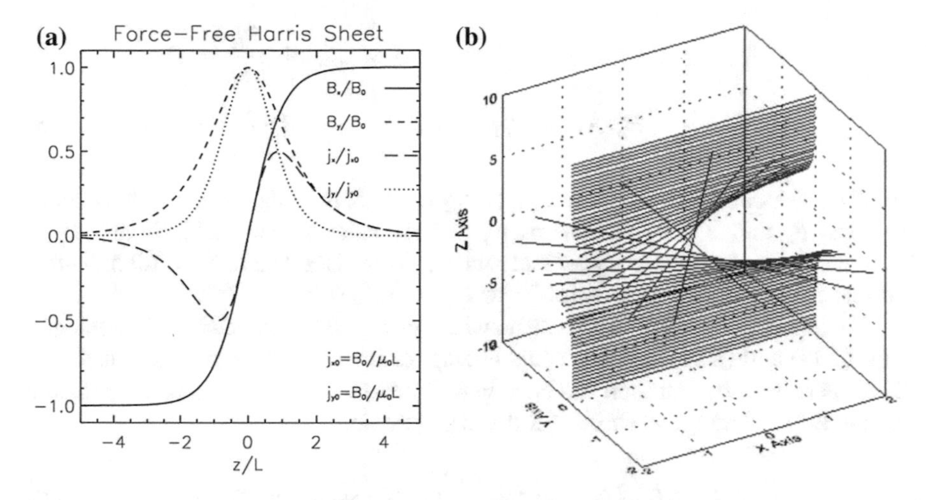

Fig. 3.3 **a** shows the magnetic field and current density components for the FFHS. **b** shows the magnetic field lines. Images copyright: M.G. Harrison's PhD thesis (Harrison 2009), (reproduced with permission)

$$\boldsymbol{B} = B_0 \left(\tanh\left(z/L\right), \operatorname{sech}\left(z/L\right), 0\right), \tag{3.12}$$

$$\boldsymbol{j} = \frac{B_0}{\mu_0 L} \frac{1}{\cosh(z/L)} \left(\tanh\left(z/L\right), \operatorname{sech}\left(z/L\right), 0\right), \tag{3.13}$$

$$P_{zz}(z) = P_T - \frac{B_0^2}{2\mu_0} = \text{const.} \tag{3.14}$$

with L the width of the current sheet, B_0 the constant magnitude of the magnetic field, $\alpha(z) = L^{-1}\operatorname{sech}(z/L)$ and P_T the total pressure. The magnetic field and current density for the FFHS are displayed in Fig. 3.3.

The DF found by Abraham-Shrauner (2013) is consistent with magnetic fields more general than the FFHS, described by Jacobi elliptic functions,

$$\boldsymbol{B} = B_0 \left(\operatorname{sn}\left(\frac{z}{L}, k\right), \operatorname{cn}\left(\frac{z}{L}, k\right), 0\right),$$

with sn and cn doubly periodic generalisations of the trigonometric functions. The parameter k is a real number such that as $k \to 0$, sn $\to$ sin and cn $\to$ cos; whereas for $k \to 1$, sn $\to$ tanh and cn $\to$ sech. As such the FFHS is a special case, as is the linear force-free case when $k \to 0$. We also note work on 'nearly' force-free equilibria (Artemyev 2011), with the FFHS modified by adding a small B_z component.

As demonstrated by Harrison and Neukirch (Harrison and Neukirch (2009a)), Neukirch et al. (Neukirch et al. (2009)), the assumption of summative separability for P_{zz} (the first option in Eq. (2.16)), determines the components of the pressure according to

$$P_{zz}(A_x, A_y) + \frac{B_0^2}{2\mu_0} = P_T,$$

$$P_1(A_x) + \frac{1}{2\mu_0} B_y^2(A_x) = P_{T1}, \quad P_2(A_y) + \frac{1}{2\mu_0} B_x^2(A_y) = P_{T2} \tag{3.15}$$

for P_{T1}, P_{T2} constants such that $P_{T1} + P_{T2} = P_T$ is the total pressure. We choose to write B_x and B_y as functions of A_y and A_x since $B_x = -dA_y/dz$ and $B_y = dA_x/dz$. In the 'particle in a potential' analogy—as discussed in Sect. 1.3.5.3—this corresponds to writing $v_x = v_x(x(t))$, and $v_y = v_y(y(t))$.

The expressions in Eq. (3.15) can now be used as the left-hand side of the integral Eqs. (2.18) and (2.19), and one could attempt to invert the Weierstrass transforms. They were used by Harrison and Neukirch (2009a) to derive a summative pressure for the FFHS. The gauge chosen for the magnetic field was

$$\mathbf{A} = B_0 L \left(2 \arctan\left(\exp\left(\frac{z}{L}\right)\right), \ln\left(\operatorname{sech}\left(\frac{z}{L}\right)\right), 0\right), \tag{3.16}$$

and as such the pressure tensor is given by

$$P_{zz} = \frac{B_0^2}{2\mu_0}\left[\frac{1}{2}\cos\left(\frac{2A_x}{B_0 L}\right) + \exp\left(\frac{2A_y}{B_0 L}\right) + b\right]. \tag{3.17}$$

The constant $b > 1/2$ contributes to a 'background' pressure consistent with a Maxwellian distribution, required for positivity. Figure 3.4 shows the P_{zz} function as defined by Eq. (3.17), with the overlaid contour delineating the 'path' followed by

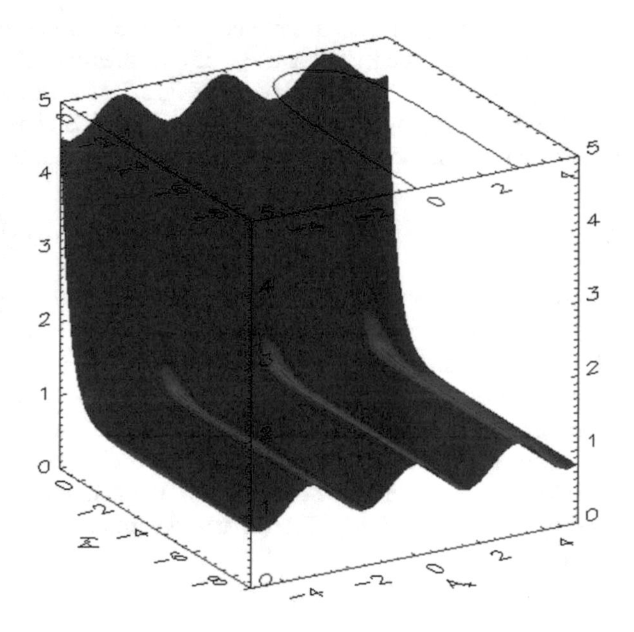

Fig. 3.4 The *Harrison-Neukirch* pressure function P_{zz}, with overlaid contour delineating the path in $(A_x(z), A_y(z))$ on which $dP_{zz}/dz = 0$. Images copyright: M.G. Harrison's PhD thesis (Harrison 2009), (reproduced with permission)

$\boldsymbol{A} = (A_x(z), A_y(z), 0)$ according to Eq. (3.16), and such that $dP_{zz}/dz = 0$. Using either Fourier transforms or inspection to invert the Weierstrass transforms, the DF calculated to correspond to the P_{zz} in Eq. (3.17) was given by

$$f_s = \frac{n_{0s}}{(\sqrt{2\pi}\, v_{\text{th},s})^3} e^{-\beta_s H_s} \left(a_s \cos\left(\beta_s u_{xs} p_{xs}\right) + e^{\beta_s u_{ys} p_{ys}} + b_s \right).$$

In this representation, u_{xs} and u_{ys} are bulk flow parameters in the x and y directions respectively, with

$$V_{xs} = \frac{u_{ys} \sinh(z/L)}{(b+1/2)\cosh^2(z/L)},$$
$$V_{ys} = \frac{u_{ys}}{(b+1/2)\cosh^2(z/L)},$$

and $|u_{xs}| = |u_{ys}|$.

3.3.1.2 Summative Pressures and the Plasma Beta

A free choice of the plasma beta is not possible in the summative *Harrison-Neukirch* equilibrium DF: it is bounded below by unity. In fact it is a feature generally observed that for pressure tensors (that correspond to force-free fields) constructed in this manner (Harrison and Neukirch 2009a; Wilson and Neukirch 2011; Abraham-Shrauner 2013; Kolotkov et al. 2015), that the plasma-beta is bounded below by unity. By combining Eqs. (3.11) and (3.15) we see that under the following assumptions,

1. $P_1(A_x) \geq 0$ and $P_2(A_y) \geq 0$
2. $\exists\, z_1, z_2$ s.t. $\sin^2 S(z_1) = 1$, $\sin^2 S(z_2) = 0$, $\cos^2 S(z_2) = 1$, $\cos^2 S(z_1) = 0$.

We justify Assumption 1. by the following argument. Whilst formally we only require the sum $P_{zz} = P_1(A_x(z)) + P_2(A_y(z)) \geq 0$ (since pressure can't be negative), we do in fact require $P_1(A_x) \geq 0$ and $P_2(A_y) \geq 0$ individually. The inverse problem defined by Eq. (1.41) ties together the dependence of P_{zz} on A_x and A_y to the dependence of the DF on p_{xs} and p_{ys} respectively. As the DF must be positive with respect to the independent variation of p_{xs} or p_{ys}, so must P_{zz} be with respect to independent variations of A_x and A_y.

Assumption 2. is trivially true in the case of a 1D linear force-free field, since $S(z)$ is a linear function of z. For the case of a nonlinear force-free field in which one of the magnetic field components goes through 0, and the other tends to 0 at $\pm\infty$, Assumption 2. will hold, and this is the case for the FFHS.

If we combine Assumptions 1. and 2., then the following inequalities will hold,

$$P_{T1} = P_1(A_x(z_1)) + \frac{B_0^2}{2\mu_0}\sin^2 S(z_1) \geq \frac{B_0^2}{2\mu_0}, \tag{3.18}$$

$$P_{T2} = P_2(A_y(z_2)) + \frac{B_0^2}{2\mu_0}\cos^2 S(z_2) \geq \frac{B_0^2}{2\mu_0}. \tag{3.19}$$

In fact, since $P_{zz}(z) = \text{const.}$, and P_{T1} and P_{T2} are independent of each other through the separation of variables, we see that the inequalities in Eqs. (3.18) and (3.19) must in fact hold true for all z. Using this knowledge, and Eq. (3.15), we conclude that

$$P_T = P_{T1} + P_{T2} \geq 2\frac{B_0^2}{2\mu_0} \implies P_1(A_x) + P_2(A_y) + \frac{B_0^2}{2\mu_0} \geq 2\frac{B_0^2}{2\mu_0},$$

and then, upon dividing through by $B_0^2/(2\mu_0)$ that

$$\beta_{pl} + 1 \geq 2 \implies \beta_{pl} \geq 1.$$

3.3.1.3 Exponential Pressure Transformation

The lower bound of unity on the β_{pl} for the DFs considered by Harrison and Neukirch (2009a), Wilson and Neukirch (2011), Abraham-Shrauner (2013), Kolotkov et a. (2015) could be considered a problem for modelling the solar corona. Formally, β_{pl} is defined as the ratio of the thermal energy density to the magnetic energy density;

$$\beta_{pl} = \sum_s \beta_{pl,s} = \frac{2\mu_0 k_B}{B_0^2}\sum_s n_s T_s, \tag{3.20}$$

for n_s and T_s the number density and temperature—of species s—respectively. In a 1D Cartesian geometry, and for a DF of the form of Eq. (1.37), the following relation holds

$$P_{zz,s} = \frac{n_s}{\beta_s} = n_s k_B T_s,$$

e.g. see Channell (1976), Harrison and Neukirch (2009b). As a result the plasma beta can be written in the more familiar form,

$$\beta_{pl} = \frac{2\mu_0 P_{zz}}{B_0^2}, \quad \text{s.t. } P_{zz} = \sum_s P_{zz,s}$$

In this chapter we take the P_{zz} used in Harrison and Neukirch (2009a), Neukirch et al. (2009), Wilson and Neukirch (2011), Kolotkov et al. (2015), which is given by Eq. (3.17), and transform it as in Eq. (3.10) with the exponential function according to

$$\psi(P_{zz}) = \exp\left[\frac{1}{P_0}\left(P_{zz} - P_{ff}\right)\right], \tag{3.21}$$

with P_0 a freely chosen positive constant. This gives $\bar{P}_{zz,ff} = P_0$, and so the plasma pressure can be as low or high as desired. Channell (1976) showed that under the assumptions used in this chapter,

$$P_{zz}(A_x, A_y) = \frac{\beta_e + \beta_i}{\beta_e \beta_i} n(A_x, A_y), \tag{3.22}$$

where $n = n_i = n_e$. Equation (3.20) then gives

$$\beta_{pl} = \frac{2\mu_0 P_{zz,ff}}{B_0^2} = \frac{2\mu_0 P_0}{B_0^2}.$$

Hence, a freely chosen P_0 corresponds directly to a freely chosen β_{pl}.

We note here that this pressure transformation can also be implicitly seen for the different linear force-free cases presented in the literature, although this connection has never been made. For example, the pressure function in Sestero (1967) (and implicitly in (Bobrova et al. 2001)) is an exponentiated version of that in Channell (1976); Attico and Pegoraro (1999). A further interesting aspect is that the momentum dependent parts of the DFs are also related to each other exponentially in the linear force-free case.

Obviously, even if integral Eq. (1.41) can be solved for the original function $P_{zz}(A_x, A_y)$ it is by no means clear that this is possible for the transformed function $\bar{P}_{zz}$. Usually one would expect that solving Eq. (1.41) for g_s is much more difficult after the transformation to $\bar{P}_{zz}$.

3.4 VM Equilibria for the Force-Free Harris Sheet: $\beta_{pl} \in (0, \infty)$

3.4.1 Calculating the DF

The pressure function in Eq. (3.17) describes $\beta_{pl} \geq 1$ regimes, and we are to transform according to Eqs. (3.10) and (3.21) in order to realise $\beta_{pl} < 1$, resulting in

$$\bar{P}_{zz} = P_0 \exp\left\{\frac{1}{2\beta_{pl}}\left[\cos\left(\frac{2A_x}{B_0 L}\right) + 2\exp\left(\frac{2A_y}{B_0 L}\right) - 1\right]\right\}.$$

The $-1/(2\beta_{pl})$ term comes from the fact that $P_{ff} = B_0^2/(2\mu_0)(1 + (b - 1/2))$, readily seen for $z = 0$, for example. Note that P_{zz} is constant over z, and so we can evaluate at any z to calculate P_{ff}. Exponentiation of P_{zz} has clearly resulted in a complicated LHS of Eq. (1.41), i.e.

$$P_0 \exp\left\{\frac{1}{2\beta_{pl}}\left[\cos\left(\frac{2A_x}{B_0L}\right)+2\exp\left(\frac{2A_y}{B_0L}\right)-1\right]\right\}=\frac{\beta_e+\beta_i}{\beta_e\beta_i}\frac{n_{0s}}{2\pi m_s^2 v_{\text{th},s}^2}$$
$$\times\int_{-\infty}^{\infty}\int_{-\infty}^{\infty} e^{-\beta_s\left((p_{xs}-q_sA_x)^2+(p_{ys}-q_sA_y)^2\right)/(2m_s)} g_s(p_{xs},p_{ys})dp_{xs}dp_{ys}, \quad (3.23)$$

and so the inverse problem defined above is mathematically challenging.

Since exponentiation of the 'summative' pressure function results in a 'multiplicative' one, we shall exploit separation of variables by assuming $g_s \propto g_{1s}(p_{xs})g_{2s}(p_{ys})$, whilst noting that $\bar{P}_{zz} \propto \bar{P}_1(A_x)\bar{P}_2(A_y)$. This assumption leads to integral equations of the form of those in Eqs. (2.18) and (2.19),

$$\bar{P}_1(A_x) \propto \int_{-\infty}^{\infty} e^{-\beta_s(p_{xs}-q_sA_x)^2/(2m_s)} g_1(p_{xs})dp_{xs}, \quad (3.24)$$

$$\bar{P}_2(A_y) \propto \int_{-\infty}^{\infty} e^{-\beta_s(p_{ys}-q_sA_y)^2/(2m_s)} g_2(p_{ys})dp_{ys}, \quad (3.25)$$

in which the LHS are formed of exponentiated cosine and exponential functions, respectively. From Eq. (3.23), we see that the inverse problem now defined by Eqs. (3.24) and (3.25) is not analytically soluble by Fourier transform methods. Hence, we resolve to use the Hermite polynomial method from Chap. 2.

The first step is to Maclaurin expand the exponentiated pressure function of Eq. (3.17) according to Eqs. (3.10) and (3.21). Exponentiation of a power series is a combinatoric problem, and was tackled by E.T. Bell in Bell (1934). If $h(x)=\exp k(x)$, and $k(x)$ is given by the power series

$$k(x)=\sum_{n=1}^{\infty}\frac{1}{n!}\zeta_n x^n,$$

then

$$h(x)=\sum_{n=0}^{\infty}\frac{1}{n!}Y_n(\zeta_1,\zeta_2,\ldots,\zeta_n)x^n,$$

for Y_n the nth Complete Bell Polynomial (CBP), with $Y_0=1$. These can be defined explicitly for $n\geq 1$ by Faà di Bruno's determinant formula as the determinant of an $n\times n$ matrix (Johnson 2002),

$$Y_n(\zeta_1,\zeta_2,\ldots\zeta_n)=\begin{vmatrix} \binom{n-1}{0}\zeta_1 & \binom{n-1}{1}\zeta_2 & \binom{n-1}{2}\zeta_3 & \cdots & \binom{n-1}{n-2}\zeta_{n-1} & \binom{n-1}{n-1}\zeta_n \\ -1 & \binom{n-2}{0}\zeta_1 & \binom{n-2}{1}\zeta_2 & \cdots & \binom{n-2}{n-3}\zeta_{n-2} & \binom{n-2}{n-1}\zeta_{n-1} \\ 0 & -1 & \binom{n-3}{0}\zeta_1 & \cdots & \binom{n-3}{n-4}\zeta_{n-3} & \binom{n-3}{n-3}\zeta_{n-2} \\ \vdots & \vdots & \vdots & & \vdots & \vdots \\ 0 & 0 & 0 & \cdots & \binom{1}{0}\zeta_1 & \binom{1}{1}\zeta_2 \\ 0 & 0 & 0 & \cdots & -1 & \binom{0}{0}\zeta_1 \end{vmatrix}. \quad (3.26)$$

For example $Y_1(\zeta_1) = \zeta_1$ and $Y_2(\zeta_1, \zeta_2) = \zeta_1^2 + \zeta_2$. We include this determinant form here since this is the representation we use to plot the DF. Instructive references on CBPs can be found in Riordan (1958), Comtet (1974), Kölbig (1994), Connon (2010), for example. Another representation for the CBPs is given by Connon (2010), where for $n \geq 1$ the Y_n can be written as

$$Y_n(\zeta_1, \zeta_2, \dots \zeta_n) = \sum_{\pi(n)} \frac{n!}{k_1! k_2! \dots k_n!} \left(\frac{\zeta_1}{1!}\right)^{k_1} \left(\frac{\zeta_2}{2!}\right)^{k_2} \dots \left(\frac{\zeta_n}{n!}\right)^{k_n}, \tag{3.27}$$

where the sum is taken over all partitions $\pi(n)$ of n, i.e. over all sets of integers k_j such that

$$k_1 + 2k_2 + \cdots + nk_n = n.$$

Using CBPs, and a simple scaling argument (Bell 1934; Connon 2010), immediately seen from Eq. (3.27),

$$Y_n(a\zeta_1, a^2\zeta_2, \dots, a^n\zeta_n) = a^n Y_n(\zeta_1, \zeta_2, \dots \zeta_n), \tag{3.28}$$

we can derive the Maclaurin expansion of the transformed pressure, making use of

$$\cos\left(\frac{2A_x}{B_0 L}\right) = \sum_{n=0}^{\infty} \frac{(-1)^n}{(2n)!} \left(\frac{2A_x}{B_0 L}\right)^{2n}, \quad \exp\left(\frac{2A_y}{B_0 L}\right) = \sum_{n=0}^{\infty} \frac{(-1)^n}{(2n+1)!} \left(\frac{2A_y}{B_0 L}\right)^{2n+1}.$$

The Maclaurin expansion is found to be

$$\bar{P}_{zz} = P_0 e^{-1/(2\beta_{pl})} \sum_{m=0}^{\infty} a_{2m} \left(\frac{A_x}{B_0 L}\right)^{2m} \sum_{n=0}^{\infty} b_n \left(\frac{A_y}{B_0 L}\right)^n, \tag{3.29}$$

with

$$a_{2m} = e^{1/(2\beta_{pl})} \frac{(-1)^m 2^{2m}}{(2m)!} Y_{2m}\left(0, \frac{1}{2\beta_{pl}}, 0, \dots, 0, \frac{1}{2\beta_{pl}}\right), \tag{3.30}$$

and

$$b_n = e^{1/\beta_{pl}} \frac{2^n}{n!} Y_n\left(\frac{1}{\beta_{pl}}, \dots, \frac{1}{\beta_{pl}}\right). \tag{3.31}$$

This allows us to formally solve the inverse problem for the unknown functions $g_{1s}(p_{xs})$ and $g_{2s}(p_{ys})$ in terms of Hermite polynomials (using results from Chap. 2), giving

$$f_s(H_s, p_{xs}, p_{ys}) = \frac{n_{0s}}{\left(\sqrt{2\pi}\, v_{\mathrm{th},s}\right)^3} e^{-1/(2\beta_{pl})} \times$$

$$\left[\sum_{m=0}^{\infty} C_{2m,s} H_{2m}\left(\frac{p_{xs}}{\sqrt{2} m_s v_{\mathrm{th},s}}\right) \sum_{n=0}^{\infty} D_{ns} H_n\left(\frac{p_{ys}}{\sqrt{2} m_s v_{\mathrm{th},s}}\right)\right] e^{-\beta_s H_s}, \quad (3.32)$$

for species-dependent coefficients $C_{2m,s}$ and D_{ns}. As discussed in Chap. 2, we fix the micro-macroscopic parameter relationships by the following conditions

$$\sigma(A_x, A_y) = 0,$$

$$P_0 \exp\left\{\frac{1}{2\beta_{pl}}\left[\cos\left(\frac{2A_x}{B_0 L}\right) + 2\exp\left(\frac{2A_y}{B_0 L}\right) - 1\right]\right\} = m_s \sum_s \int v_z^2 f_s d^3 v,$$

for the f_s given by Eq. (3.32). After performing the necessary integrations, these conditions are satisfied by fixing the parameters according to

$$n_{0i} = n_{0e} = n_0, \quad P_0 = n_0 \frac{\beta_e + \beta_i}{\beta_e \beta_i}$$

$$C_{2m,s} = \left(\frac{\delta_s}{\sqrt{2}}\right)^{2m} a_{2m}, \quad D_{ns} = \mathrm{sgn}(q_s)^n \left(\frac{\delta_s}{\sqrt{2}}\right)^n b_n.$$

As yet, the distribution of Eq. (3.32), together with the micro-macroscopic conditions, is only a formal solution to the inverse problem posed, and we now proceed to confirm the convergence and boundedness properties, using techniques from Chap. 2.

3.4.2 *Convergence and Boundedness of the DF*

Here we include the full details of the calculations that confirm the validity of the Hermite Polynomial representation of the multiplicative FFHS equilibrium in the 'original' gauge (Eq. (3.16)). We shall first verify the convergence of g_{2s} (expanded over n in Eq. (3.32)) using the convergence condition from Sect. 2.4, and then verify convergence of g_{1s} by comparison with g_{2s}.

3.4.2.1 Convergence of the p_{ys} Dependent Sum

As Theorem 1 states, we can verify convergence of g_{2s} provided

$$\lim_{n\to\infty} \sqrt{n}\left|\frac{b_{n+1}}{b_n}\right| < 1/\delta_s.$$

Explicit expansion of the exponentiated exponential series by 'twice' using Maclaurin series (as opposed to the CBP formulation of Eq. (3.31)) gives

$$\begin{aligned}
\tilde{P}_2(\tilde{A}_y) = \exp\left(\frac{1}{\beta_{pl}}\exp\left(\frac{2A_y}{B_0 L}\right)\right) &= \sum_{k=0}^{\infty}\frac{1}{\beta_{pl}^k k!}\exp\left(\frac{2kA_y}{B_0 L}\right),\\
&= \sum_{k=0}^{\infty}\frac{1}{\beta_{pl}^k k!}\sum_{n=0}^{\infty}\frac{2^n k^n}{n!}\left(\frac{A_y}{B_0 L}\right)^n\\
&= \sum_{n=0}^{\infty} b_n\left(\frac{A_y}{B_0 L}\right)^n,
\end{aligned}$$

such that b_n are defined by

$$b_n = \frac{2^n}{n!}\sum_{k=0}^{\infty}\frac{k^n}{\beta_{pl}^k k!}, \tag{3.33}$$

And for which the sum over k is itself a convergent series, meaning that the b_n are well-defined. Using the definition of b_n and b_{n+1} gives

$$\begin{aligned}
b_{n+1}/b_n &= \frac{2}{n+1}\sum_{j=0}^{\infty}\frac{j^{n+1}}{j!\beta_{pl}^j}\Bigg/\sum_{j=0}^{\infty}\frac{j^n}{j!\beta_{pl}^j}\\
&= \frac{2}{n+1}\left(\frac{0+\frac{1}{0!\beta_{pl}}+\frac{2^n}{1!\beta_{pl}^2}+\frac{3^n}{2!\beta_{pl}^3}+\cdots}{0+\frac{1}{1!\beta_{pl}}+\frac{2^n}{2!\beta_{pl}^2}+\frac{3^n}{3!\beta_{pl}^3}+\cdots}\right)\\
&= \frac{2}{n+1}\left(\frac{\frac{1}{\beta_{pl}}+2\frac{2^n}{2!\beta_{pl}^2}+3\frac{3^n}{3!\beta_{pl}^3}+\cdots}{\frac{1}{1!\beta_{pl}}+\frac{2^n}{2!\beta_{pl}^2}+\frac{3^n}{3!\beta_{pl}^3}+\cdots}\right).
\end{aligned}$$

The kth 'partial sum' of this fraction has the form

$$\mathcal{S}_{n,k} = \frac{p_1 + 2p_2 + 3p_3 + \cdots + kp_k}{p_1 + p_2 + p_3 + \cdots}$$

with $p_i \asymp 1/i!$, where we write $g \asymp h$ to mean g/h and h/g are bounded away from 0. Now since the denominator of the p_i increase factorially we have $ip_i \asymp p_i$ and hence

$$0 < \sum_{i=1}^{\infty} ip_i < \infty \quad \text{and} \quad 0 < \sum_{i=1}^{\infty} p_i < \infty.$$

Thus $\mathcal{S}_{n,k} \to \mathcal{S}_{n,\infty} \in (0, \infty)$ and, more specifically, $\mathcal{S}_{n,\infty} \asymp 1$ in n. Therefore

$$b_{n+1}/b_n = \mathcal{S}_{n,\infty}/(n+1) \asymp 1/n.$$

That is to say b_{n+1}/b_n behaves asymptotically like $1/n$. This satisfies the condition of Theorem 1. Hence $g_{2s}(p_{ys})$ converges for all δ_s and p_{ys} by the comparison test.

3.4.2.2 Convergence of the p_{xs} Dependent Sum

We shall now verify convergence of g_{1s}, by comparison with g_{2s}. By explicitly using the Maclaurin expansion of the exponential, and then the power-series representation for $\cos^n x$ from Gradshteyn and Ryzhik (2007)

$$\cos^{2n} x = \frac{1}{2^{2n}}\left[\sum_{k=0}^{n-1} 2\binom{2n}{k}\cos(2(n-k)x) + \binom{2n}{n}\right],$$

$$\cos^{2n-1} x = \frac{1}{2^{2n-2}}\sum_{k=0}^{n-1}\binom{2n-1}{k}\cos((2n-2k-1)x),$$

one can calculate

$$\tilde{P}_1(\tilde{A}_x) = \exp\left(\frac{1}{2\beta_{pl}}\cos\left(\frac{2A_x}{B_0 L}\right)\right) = \sum_{m=0}^{\infty} a_{2m}\left(\frac{A_x}{B_0 L}\right)^{2m}.$$

The zeroth coefficient is given by $a_0 = \exp\left(1/(2\beta_{pl})\right)$, and the rest are

$$a_{2m} = \frac{2(-1)^m}{(2m)!}\sum_{k=0}^{\infty}\sum_{j\in J_k}\frac{1}{j!(4\beta_{pl})^j}\binom{j}{k}(j-2k)^{2m},$$

for $J_k = \{2k+1, 2k+2, \ldots\}$ and $m \neq 0$. By rearranging the order of summation, a_{2m} can be written

$$a_{2m} = \frac{2(-1)^m}{(2m)!}\sum_{j=1}^{\infty}\frac{1}{j!(4\beta_{pl})^j}\sum_{k=0}^{\lfloor (j-1)/2\rfloor}\binom{j}{k}(j-2k)^{2m},$$

where $\lfloor x \rfloor$ is the floor function, denoting the greatest integer less than or equal to x. Recognising an upper bound in the expression for a_{2m};

$$\sum_{n=0}^{\lfloor (j-1)/2\rfloor}\binom{j}{n}(j-2n)^{2m} \leq j^{2m}\sum_{n=0}^{j}\binom{j}{n} = 2^j j^{2m},$$

gives

$$
\begin{aligned}
a_{2m} < \frac{2(-1)^m}{(2m)!}\sum_{j=1}^{\infty}\frac{2^{j+1}j^{2m}}{j!2^j(2\beta_{pl})^j} &= 2\frac{(-1)^m}{(2m)!}\sum_{j=1}^{\infty}\frac{j^{2m}}{j!(2\beta_{pl})^j},\\
&\leq \frac{2}{(2m)!}\sum_{j=1}^{\infty}\frac{j^{2m}}{j!(2\beta_{pl})^j},\\
&= \frac{1}{(2m)!}\sum_{j=1}^{\infty}\frac{2^{1-j}j^{2m}}{j!\beta_{pl}^j} < b_{2m}.
\end{aligned}
$$

Hence we now have an upper bound on a_{2m} for $m \neq 0$ and we know that $a_{2m+1} = 0$, and so is bounded above by b_{2m+1}. Note also that $a_0 < b_0$. Hence, each term in our series for $g_{1s}(p_{xs})$ is bounded above by a series known to converge for all δ_s according to

$$
a_l\left(\frac{\delta_s}{\sqrt{2}}\right)^l H_l(x) < b_l\left(\frac{\delta_s}{\sqrt{2}}\right)^l H_l(x), \quad \forall l.
$$

So by the comparison test, we can now say that $g_{1s}(p_{xs})$ is a convergent series. Hence the representation of the DF in Eq. (3.32) is convergent.

3.4.2.3 Boundedness of the DF

The boundedness of the DF in Eq. (3.32) is now guaranteed by the reasoning from Sect. 2.4.1.2 for a general solution, and need not be repeated here.

3.4.3 Moments of the DF

The moments of the DF are used to calculate the number density and bulk velocity, and in turn the charge and current densities respectively. It is useful to calculate these quantities from the DF to confirm parity with the required macroscopic quantities not only as a procedural check, but also to derive relations between the micro- and macroscopic parameters.

3.4.3.1 The Zeroth Order Moment

The number density is found by taking the zeroth moment;

$$n_s(A_x, A_y) = \frac{e^{-\frac{1}{2\beta_{pl}}}}{m_s^3} \frac{n_0}{(\sqrt{2\pi}\, v_{\mathrm{th},s})^3} \times$$
$$\left[\sum_{m=0}^{\infty} C_{2m,s} \int_{-\infty}^{\infty} e^{-\frac{\beta_s}{2m_s}(p_{xs}-q_s A_x)^2} H_{2m}\left(\frac{p_{xs}}{\sqrt{2}m_s v_{\mathrm{th},s}}\right) dp_{xs} \times \right.$$
$$\left. \sum_{n=0}^{\infty} D_{ns} \int_{-\infty}^{\infty} e^{-\frac{\beta_s}{2m_s}(p_{ys}-q_s A_y)^2} H_n\left(\frac{p_{ys}}{\sqrt{2}m_s v_{\mathrm{th},s}}\right) dp_{ys} \int_{-\infty}^{\infty} e^{-\frac{\beta_s}{2m_s}p_{zs}^2} dp_{zs} \right],$$

which, after integrating over p_{zs} and making substitutions, gives

$$n_s(A_x, A_y) = \frac{n_0 e^{-\frac{1}{2\beta_{pl}}}}{\pi} \sum_{m=0}^{\infty} C_{2m,s} \int_{-\infty}^{\infty} e^{-(X - \frac{q_s A_x}{\sqrt{2}m_s v_{\mathrm{th},s}})^2} H_{2m}(X) dX$$
$$\times \sum_{n=0}^{\infty} D_{ns} \int_{-\infty}^{\infty} e^{-(Y - \frac{q_s A_y}{\sqrt{2}m_s v_{\mathrm{th},s}})^2} H_n(Y) dY.$$

Use the standard integral (Gradshteyn and Ryzhik 2007),

$$\int_{-\infty}^{\infty} e^{-(x-y)^2} H_n(x) dx = \sqrt{\pi} 2^n y^n,$$

to give

$$n_s(A_x, A_y) = n_0 e^{-\frac{1}{2\beta_{pl}}} \sum_{m=0}^{\infty} C_{2m,s} 2^{2m} \left(\frac{q_s A_x}{\sqrt{2}m_s v_{\mathrm{th},s}}\right)^{2m} \sum_{n=0}^{\infty} D_{ns} 2^n \left(\frac{q_s A_y}{\sqrt{2}m_s v_{\mathrm{th},s}}\right)^n$$
$$= \frac{n_0}{P_0} \bar{P}_{zz}.$$

Using $P_{zz,ff} = P_0$, we see that

$$n_{ff} = n_0,$$

and so n_0 represents the constant particle number density.

3.4.3.2 The v_x Moment

We now take the first moment of the DF by v_x denoted by $[v_x f_s]$;

$$
\begin{aligned}
[v_x f_s] &= \frac{1}{m_s^3}\int_{-\infty}^{\infty}\int_{-\infty}^{\infty}\int_{-\infty}^{\infty} v_x f_s d^3p, \\
&= \frac{n_0 e^{-\frac{1}{2\beta_{pl}}}}{(\sqrt{2\pi})m_s v_{\mathrm{th},s}} \sum_{n=0}^{\infty} b_n \left(\frac{A_y}{B_0 L}\right)^n \sum_{m=0}^{\infty} C_{2m,s} \times \\
&\underbrace{\int_{-\infty}^{\infty} v_x e^{-\frac{\beta_s}{2m_s}(p_{xs}-q_sA_x)^2} H_{2m}\left(\frac{p_{xs}}{\sqrt{2}m_s v_{\mathrm{th},s}}\right) dp_{xs}}_{I_{vx}},
\end{aligned}
$$

after both the p_{ys} and p_{zs} integrations. Now, use the Hermite expansion of the exponential (Morse and Feshbach 1953), to give

$$
\begin{aligned}
I_{v_x} = &\frac{1}{m_s}\int_{-\infty}^{\infty}(p_{xs} - q_s A_x) H_{2m}\left(\frac{p_{xs}}{\sqrt{2}m_s v_{\mathrm{th},s}}\right) e^{-\frac{\beta_s p_{xs}^2}{2m_s}} \times \\
&\left[\sum_{j=0}^{\infty}\frac{1}{(j)!}\left(\frac{q_s A_x}{\sqrt{2}m_s v_{\mathrm{th},s}}\right)^j H_j\left(\frac{p_{xs}}{\sqrt{2}m_s v_{\mathrm{th},s}}\right)\right] dp_{xs}.
\end{aligned}
$$

Now define an inner product according to

$$
\langle f_1(x), f_2(x)\rangle = \int_{-\infty}^{\infty} e^{-x^2} f_1(x) f_2(x) dx. \tag{3.34}
$$

Then orthogonality of the Hermite polynomials (Eq. (2.3)), and the recurrence relation, $H_{n+1}(x) = 2xH_n(x) - 2nH_{n-1}(x)$, are used to give

$$
\begin{aligned}
\langle xH_j(x), H_{2m}(x)\rangle &= j\langle H_{j-1}(x), H_{2m}(x)\rangle + \frac{1}{2}\langle H_{j+1}(x), H_{2m}(x)\rangle \\
&= \sqrt{\pi}2^{2m}(2m)!\left(j\delta_{j-1,2m} + \frac{1}{2}\delta_{j+1,2m}\right).
\end{aligned} \tag{3.35}
$$

This allows us to write

$$
\begin{aligned}
I_{v_x} = &\sqrt{2\pi}v_{\mathrm{th},s}2^{2m}(2m)! \times \\
&\sum_{j=0}^{\infty}\frac{1}{j!}\left(\frac{q_s A_x}{\sqrt{2}m_s v_{\mathrm{th},s}}\right)^j\left[\sqrt{2}m_s v_{\mathrm{th},s}\left(j\delta_{j-1,2m} + \frac{1}{2}\delta_{j+1,2m}\right) - q_s A_x \delta_{j,2m}\right].
\end{aligned}
$$

Hence, we have

$$[v_x f_s] = \frac{n_0 e^{-\frac{1}{2\beta_{pl}}}}{m_s} \sum_{n=0}^{\infty} b_n \left(\frac{A_y}{B_0 L}\right)^n \sum_{m=0}^{\infty} C_{2m,s} 2^{2m} (2m)!$$
$$\times \sum_{j=0}^{\infty} \frac{1}{j!} \left(\frac{q_s A_x}{\sqrt{2} m_s v_{\text{th},s}}\right)^j \left[\sqrt{2} m_s v_{\text{th},s} \left(j\delta_{j-1,2m} + \frac{1}{2}\delta_{j+1,2m}\right) - q_s A_x \delta_{j,2m}\right].$$

reducing to

$$[v_x f_s] = \left(\frac{m_s v_{\text{th},s}^2}{q_s B_0 L}\right) n_0 e^{-\frac{1}{2\beta_{pl}}} \sum_{n=0}^{\infty} b_n \left(\frac{A_y}{B_0 L}\right)^n \sum_{m=1}^{\infty} a_{2m} 2m \left(\frac{A_x}{B_0 L}\right)^{2m-1}$$
$$= \left(\frac{m_s v_{\text{th},s}^2}{q_s P_0}\right) n_0 \frac{\partial \bar{P}_{zz}}{\partial A_x} = \frac{\beta_e \beta_i}{\beta_e + \beta_i} \left(\frac{1}{q_s \beta_s}\right) \frac{\partial \bar{P}_{zz}}{\partial A_x} \tag{3.36}$$

The x component of current density is defined as $j_x = \sum_s q_s [v_x f_s]$, giving

$$j_x = \frac{\beta_e \beta_i}{\beta_e + \beta_i} \frac{\partial \bar{P}_{zz}}{\partial A_x} \sum_s \frac{1}{\beta_s} = \frac{\partial \bar{P}_{zz}}{\partial A_x}$$
$$\implies j_x = \frac{\partial \bar{P}_{zz}}{\partial A_x}, \tag{3.37}$$

reproducing the familiar result from e.g. Channell (1976), Harrison and Neukirch (2009b), Schindler (2007), Mynick et al. (1979). The first moment of the DF can also be used to calculate the bulk velocity in terms of the microscopic parameters;

$$V_{xs} = \frac{[v_x f_s]}{n_s} = \frac{j_x}{q_s \beta_s P_0}, \tag{3.38}$$

using Eq. (3.36). Then, by using the current density for the FFHS (Eq. (3.13)),

$$\boldsymbol{j} = \frac{B_0}{\mu_0 L} \left(\frac{\sinh\left(\frac{z}{L}\right)}{\cosh^2\left(\frac{z}{L}\right)}, \frac{1}{\cosh^2\left(\frac{z}{L}\right)}, 0\right), \tag{3.39}$$

we have the bulk flow in x

$$V_{xs} = \frac{B_0}{\mu_0 L q_s \beta_s P_0} \frac{\sinh\left(\frac{z}{L}\right)}{\cosh^2\left(\frac{z}{L}\right)}. \tag{3.40}$$

3.4.3.3 The v_y Moment

By a completely analogous calculation, we derive the v_y moment of the DF,

$$
\begin{aligned}
[v_y f_s] &= \left(\frac{m_s v_{\text{th},s}^2}{P_0 q_s}\right) n_0 \frac{\partial \bar{P}_{zz}}{\partial A_y} \\
&= \frac{\beta_e \beta_i}{\beta_e + \beta_i}\left(\frac{m_s v_{\text{th},s}^2}{q_s}\right)\frac{\partial \bar{P}_{zz}}{\partial A_y}
\end{aligned}
$$

Again, the current density $j_y = \sum_s q_s [v_y f_s]$ gives

$$
\begin{aligned}
j_y &= \frac{\beta_e \beta_i}{\beta_e + \beta_i}\frac{\partial P_{zz}}{\partial A_y}\sum_s m_s v_{\text{th},s}^2 = \frac{\partial \bar{P}_{zz}}{\partial A_y} \\
\implies j_y &= \frac{\partial \bar{P}_{zz}}{\partial A_y}.
\end{aligned}
$$

We can also calculate the bulk velocity in terms of the microscopic parameters;

$$
V_{ys} = \frac{B_0}{\mu_0 L q_s \beta_s P_0}\frac{1}{\cosh^2(z/L)}. \tag{3.41}
$$

3.4.4 *Properties of the DF*

3.4.4.1 Current Sheet Width

The nature of the inverse problem is to calculate a microscopic description of a system, given certain prescribed macroscopic data. Hence, one of the main tasks is to find the relationships between the characteristic parameters of each level of description. That is to say, given (B_0, P_0, L) for example, what is their relation to $(m_s, q_s, v_{\text{th},s}, n_{0s})$?

Currently, there are six free parameters that will determine the nature of the equilibrium. These are n_0, β_{pl}, $\beta_{th,i}$, $\beta_{th,e}$, δ_i and δ_e. n_0 is in principle fixed by ensuring that the DF is normalised to the total particle number. As yet we have no information regarding the width of the current sheet L. To this end we shall consider bulk velocities V_{xs} and V_{ys}, obtained from the first moment of the DF. The calculations in Sect. 3.4.3, together with the fact that $B_0 = \sqrt{2\mu_0 P_0/\beta_{pl}}$ give

$$
\begin{aligned}
V_{xs} &= \frac{[v_x f_s]}{n_0} = \sqrt{\frac{2}{\mu_0 \beta_{pl} P_0}}\frac{1}{L q_s \beta_s}\frac{\sinh(z/L)}{\cosh^2(z/L)}, \\
V_{ys} &= \frac{[v_y f_s]}{n_0} = \sqrt{\frac{2}{\mu_0 \beta_{pl} P_0}}\frac{1}{L q_s \beta_s}\frac{1}{\cosh^2(z/L)}.
\end{aligned}
$$

We can identify the coefficient of the z dependent profiles as the amplitude of the bulk velocities, V_{xs} and V_{ys}, as u_s, given by

$$u_s = \sqrt{\frac{2}{\mu_0 \beta_{pl} P_0} \frac{1}{L q_s \beta_s}}, \tag{3.42}$$

giving

$$(u_i - u_e)^2 = \frac{2}{\mu_0 \beta_{pl} P_0 L^2 e^2} \left(\frac{\beta_e + \beta_i}{\beta_e \beta_i} \right)^2, \tag{3.43}$$

$$\implies L = \frac{1}{e} \sqrt{\frac{2(\beta_e + \beta_i)}{\mu_0 n_0 \beta_e \beta_i (u_i - u_e)^2 \beta_{pl}}}, \tag{3.44}$$

where $e = |q_s|$. Interestingly, this is almost identical to the expression found in Neukirch et al. (2009) for the current sheet width of the Harrison-Neukirch equilibrium, with the addition of the $\beta_{pl}^{1/2}$ factor in the denominator. It is readily seen that, given some fixed B_0, $L \propto \beta_{pl}^{-1/2}$. This makes sense in that, by raising the number density n_0, and hence β_{pl}, there are simply more current carriers available to produce $\boldsymbol{j}$, and hence the width L can reduce. By manipulating Eq. (3.42) one can show that the amplitudes of the fluid velocities are given by

$$\frac{u_s}{v_{\mathrm{th},s}} = 2\mathrm{sgn}(q_s) \frac{\delta_s}{\beta_{pl}} = 2\mathrm{sgn}(q_s) \frac{\rho_s}{L \beta_{pl}}. \tag{3.45}$$

Once again, this is almost identical to the expression found in Neukirch et al. (2009), with the addition of a β_{pl} factor in the denominator.

3.4.4.2 Plots of the DF

Having found mathematical expressions for the DFs, we now present different plots of their dependence on v_x and v_y, for $z/L = 0, -1, 1$. Plotting f_s in the original gauge is a challenging numerical task, and particularly for the low-β_{pl} regime. The reasoning is as follows. When $\beta_{pl} < 1/2$, the $C_{2m,s}$ (for example) are readily seen to be of the order

$$\left(\frac{1}{\sqrt{2}} \right)^{2m} \frac{1}{(2m)!} \left(\frac{\delta_s}{\beta_{pl}} \right)^{2m},$$

since Y_{2m} is a polynomial of order $2m$ in $1/(2\beta_{pl})$. The factorial dependence in the denominator ensures that these terms $\to 0$ as $m \to \infty$. But, for relatively small m there is a competition between the factorial and the β_{pl}^{-2m}, factor. This means that one must go to many terms in the expansion to get near numerical convergence.

As a result, one needs to calculate both incredibly small (e.g. the $1/(2m)!$ factor), and incredibly large numbers (the Y_{2m} factors), and combine them to reach $C_{2m,s}$.

Furthermore, the Hermite polynomials become very large when the modulus of the argument is large. In normalised parameters, suitable for numerical methods, we have that

$$H_n\left(\frac{p_{js}}{\sqrt{2}m_s v_{\mathrm{th},s}}\right) = H_n\left(\frac{1}{\sqrt{2}}\left(\tilde{v}_{js} + \mathrm{sgn}(q_s)\delta_s^{-1}\tilde{A}_j\right)\right), \tag{3.46}$$

for $\tilde{p}_{js} = p_{js}/(m_s v_{\mathrm{th},s})$, and $\tilde{A}_j = A_j/(B_0 L)$. In particular, small values of δ_s mean that one needs to calculate H_n of a large number, which can itself be inordinately large since H_n is a polynomial.

So, while it has been proven that the series with which we represent the DFs are convergent for all values of the relevant parameters, attaining numerical convergence is difficult for the low-β_{pl} regime, and particularly for the p_{xs} dependent sum. Here we present plots for $\beta_{pl} = 0.85$ and $\delta_i = \delta_e = 0.15$. As aforementioned we use Faà di Bruno's determinant formula in Eq. (3.26) to calculate the CBP's, and a recurrence relation for the Hermite Polynomials. Whilst this β_{pl} is only modestly below unity, however it represents a value of which we are confident of our numerics for both the p_{xs} and p_{ys} dependent sums. In Figs. 3.5a–c we plot the v_x variation of our electron DF, as a representative example (the v_y plots are qualitatively similar). First of all we note that the DFs appear to have only a single maximum, and fall off as $v_x \to \pm\infty$. This is to be contrasted with the plots of the DF using the additive pressure, which can have multiple peaks (Neukirch et al. 2009). Thus far we have not found any indication of multiple peaks in the parameter regime that we have been able to explore. However, this does not mean that multiple peaks can not appear, for example for lower values of the β_{pl}.

A first look at the plots also seems to indicate that the shape of the DF resembles the shape of a Maxwellian. Motivated by this similarity, we define a Maxwellian DF according to Eq. (1.26), and repeated here,

$$f_{Maxw,s} = \frac{n_0}{(\sqrt{2\pi}\, v_{\mathrm{th},s})^3} \exp\left[\frac{-(\boldsymbol{v} - \boldsymbol{V}_s(z))^2}{2v_{\mathrm{th},s}^2}\right]. \tag{3.47}$$

The Maxwellian distribution reproduces the same first order moment in terms of z as the equilibrium solution does, namely $\boldsymbol{V}_s$, and a spatially uniform number density, namely n_0. However it is not a solution of the Vlasov equation and hence not an equilibrium solution. PIC simulations for a force-free field were initiated with a distribution of this type in Hesse et al. (2005), Birn and Hesse (2010), for example. To highlight the difference between the two DFs, we plot both the v_x and v_y variation of the ratio of the DF, with the Maxwellian of Eq. (3.47) for both ions and electrons in Figs. 3.6a–c. As we can see, in all plots the ratio deviates from unity, and in some cases these deviations are substantial. This shows that the initial impression is somewhat misleading. We also observe a symmetry in that the v_y dependent plots are even in z, since A_y and $\langle v_y\rangle_s$ are even in z.

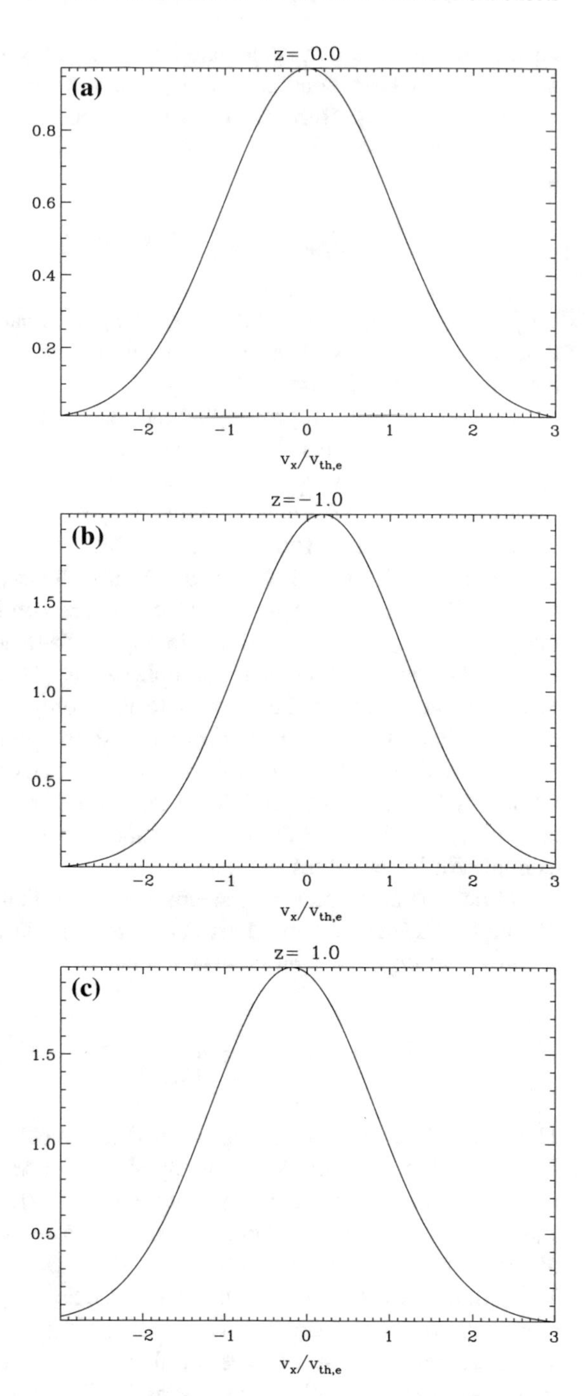

Fig. 3.5 The v_x variation of f_e for $z/L = 0$ Fig. 3.5a, $z/L = -1$ Fig. 3.5b and $z/L = 1$ Fig. 3.5c. $\beta_{pl} = 0.85$ and $\delta_e = 0.15$. Note the antisymmetry of the $z = \pm 1$ plots with respect to each other

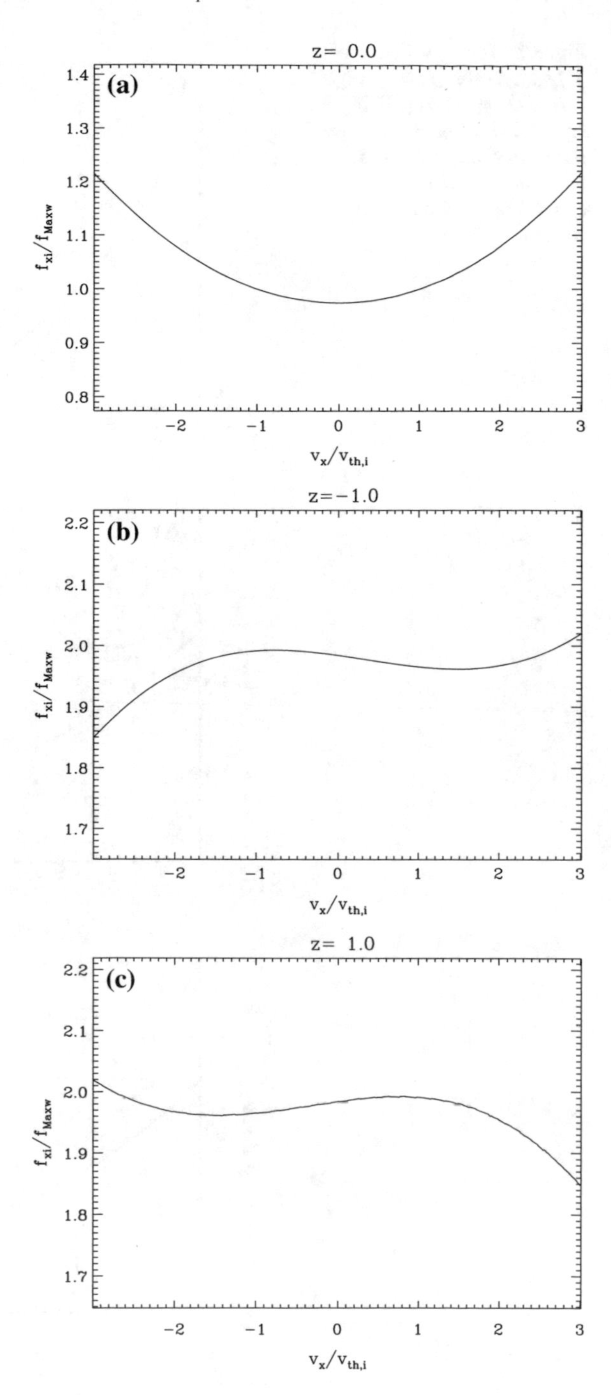

Fig. 3.6 The v_x variation of $f_i/f_{Maxw,i}$ for $z/L = 0$ Fig. 3.6a, $z/L = -1$ Fig. 3.6b and $z/L = 1$ Fig. 3.6c. $\beta_{pl} = 0.85$ and $\delta_i = 0.15$. Note the antisymmetry of the $z = \pm 1$ plots with respect to each other

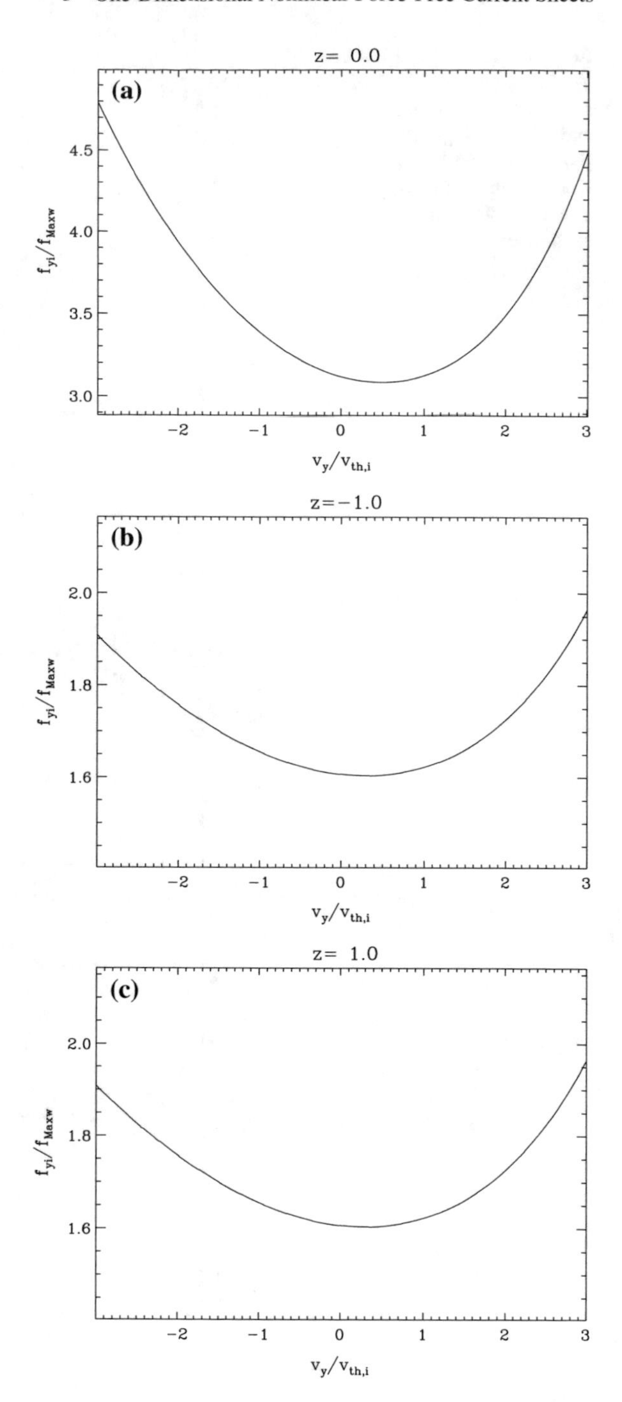

Fig. 3.7 The v_y variation of $f_i/f_{Maxw,i}$ for $z/L = 0$ Fig. 3.7a, $z/L = -1$ Fig. 3.7b and $z/L = 1$ Fig. 3.7c. $\beta_{pl} = 0.85$ and $\delta_i = 0.15$. Note the symmetry of the $z = \pm 1$ plots with respect to each other

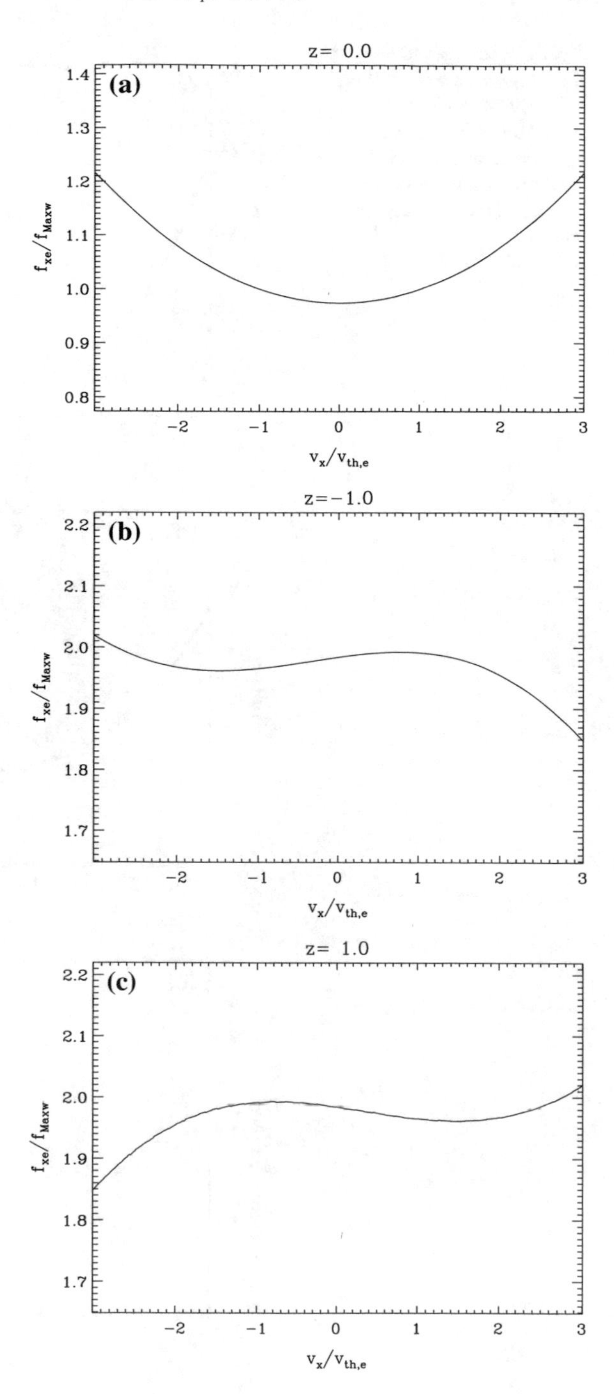

Fig. 3.8 The v_x variation of $f_e/f_{Maxw,e}$ for $z/L = 0$ Fig. 3.8a, $z/L = -1$ Fig. 3.8b and $z/L = 1$ Fig. 3.8c. $\beta_{pl} = 0.85$ and $\delta_e = 0.15$. Note the antisymmetry of the $z = \pm 1$ plots with respect to each other

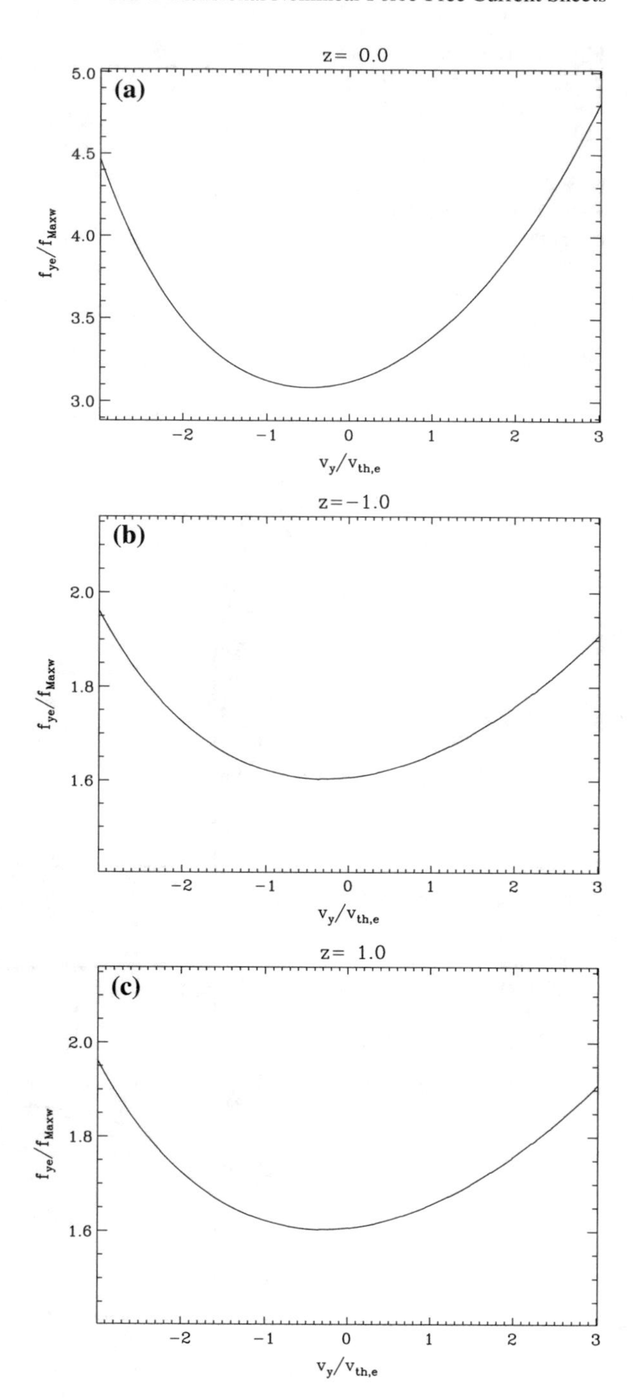

Fig. 3.9 The v_y variation of $f_e/f_{Maxw,e}$ for $z/L = 0$ Fig. 3.9a, $z/L = -1$ Fig. 3.9b and $z/L = 1$ Fig. 3.9c. $\beta_{pl} = 0.85$ and $\delta_e = 0.15$. Note the symmetry of the $z = \pm 1$ plots with respect to each other

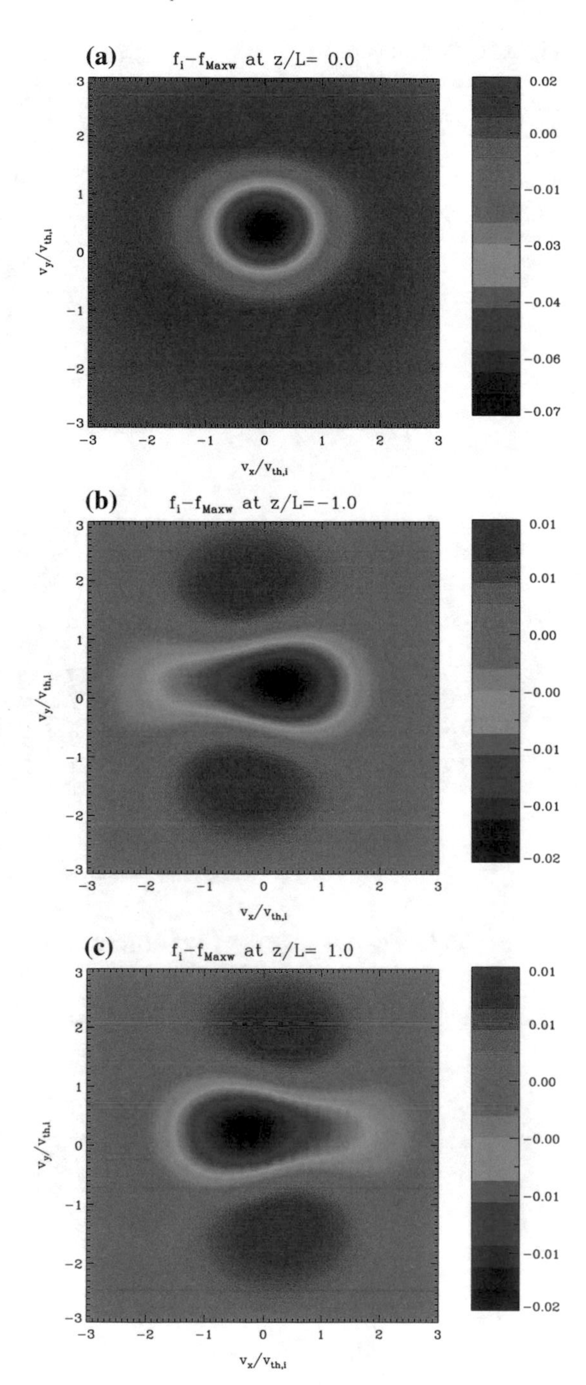

Fig. 3.10 Contour plots of $f_i - f_{Maxw,i}$ for $z/L = 0$ Fig. 3.10a, $z/L = -1$ Fig. 3.10b and $z/L = 1$ Fig. 3.10c. $\beta_{pl} = 0.85$ and $\delta_i = 0.15$. Note the antisymmetry of the $z = \pm 1$ plots with respect to each other

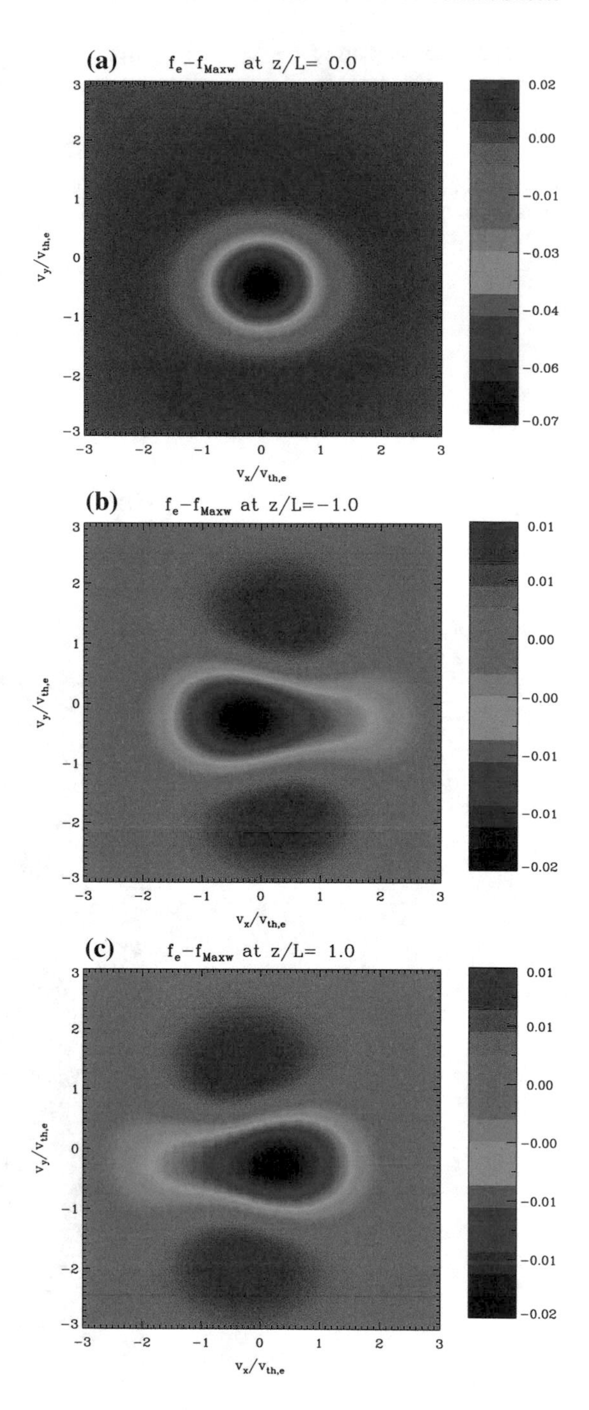

Fig. 3.11 Contour plots of $f_e - f_{Maxw,e}$ for $z/L = 0$ Fig. 3.11a, $z/L = -1$ Fig. 3.11b and $z/L = 1$ Fig. 3.11c. $\beta_{pl} = 0.85$ and $\delta_e = 0.15$. Note the antisymmetry of the $z = \pm 1$ plots with respect to each other

To further see the deviations of f_s from the Maxwellian, we present contour plots of the difference $f_s - f_{Maxw,s}$ in Figs. 3.10a–c over (v_x, v_y) space for various z values. One observation we can make from these is that there is a symmetry with respect to both velocity direction and the value of z. For example it seems that f_s is symmetric under the transformation $(v_x \to -v_x,\ z \to -z)$. This seems reasonable since A_x is dynamically equivalent to an odd function of z, by a gauge transformation, as B_y is even (more on this in Sect. 3.5.1). For a plasma-beta modestly below unity, and thermal Larmor radius roughly 15% of the current sheet width, we find distributions that are roughly Maxwellian in shape, but 'shallower' at the centre of the sheet. At the outer edges of the sheet, this shallowness assumes a drop-shaped depression in the v_x direction, with localised differences for large v_y.

3.5 'Re-Gauged' Equilibrium DF for the FFHS

3.5.1 On the Gauge for the Vector Potential

In Sect. 3.4 we used the pressure transformation techniques to derive a pressure tensor of 'multiplicative form'

$$P_{zz} = P_1(A_x)P_2(A_y),$$

in order to construct a DF self-consistent with any value of the β_{pl}. However, the exact form of the DF was challenging to calculate numerically for low β_{pl}, with plots for β_{pl} only modestly below unity presented ($\beta_{pl} = 0.85$). The 'problem terms' are those that depend on p_{xs}. The specific problem is that the A_x function in the original gauge is neither even or odd,

$$A_x = 2B_0 L \arctan\left(\exp\left(\frac{z}{L}\right)\right),$$

and as a result the range of p_{xs} for which it is necessary to numerically calculate a convergent DF can be obstructive, say over a symmetric range in velocity space. Equation (3.46) shows us that when A_x is neither even nor odd, then $|p_{xs}|$ can take on larger than 'necessary' values for a given v_x.

In this chapter, we shall 're-gauge' the vector potential component A_x to be an odd function,

$$A_x = 2B_0 L \arctan\left(\tanh\left(\frac{z}{2L}\right)\right), \tag{3.48}$$

which is commensurate with B_y being an even function and results in the same $B_y = B_0 \operatorname{sech}(z/L)$ as the one derived from the A_x defined in (3.16). As a consequence the numerical calculation of the DFs that we shall calculate for the FFHS becomes easier in the low β_{pl} regime.

3.5.2 DF for the 'Re-Gauged' FFHS: $\beta_{pl} \in (0, \infty)$

We will now calculate a multiplicative DF for the 're-gauged' FFHS, in the same style as in Sect. 3.4, in the effort to produce a low-beta DF for the FFHS that is easier to calculate numerically, and plot. The new gauge is defined by

$$\boldsymbol{A} = B_0 L \left(2 \arctan\left(\tanh\left(\frac{z}{2L}\right)\right), \ln \operatorname{sech}\frac{z}{L}, 0\right). \tag{3.49}$$

This re-gauging is equivalent to adding a constant to A_x and so corresponds to a shift in the origin of the A_x dependent part of the summative P_{zz} used in Harrison and Neukirch (2009a). As a result, one can derive a new summative pressure function in the same manner as in (Harrison and Neukirch 2009a), corresponding to this new gauge, as

$$P_{zz} = \frac{B_0^2}{2\mu_0}\left[\sin^2\left(\frac{A_x}{B_0 L}\right) + \exp\left(\frac{2A_y}{B_0 L}\right)\right] \tag{3.50}$$

The next step is to construct a multiplicative pressure tensor. Using the same pressure transformation technique as in Sect. 3.3.1.3, on the P_{zz} given in Eq. (3.50), we arrive at the 're-gauged' multiplicative pressure

$$P_{zz} = P_0 e^{-1/\beta_{pl}} \exp\left[\frac{1}{\beta_{pl}}\left(\sin^2\left(\frac{A_x}{B_0 L}\right) + \exp\left(\frac{2A_y}{B_0 L}\right)\right)\right] \tag{3.51}$$

$$= P_0 \exp\left[\sum_{n=1}^{\infty}\frac{1}{(2n)!}\nu_{2n}\left(\frac{A_x}{B_0 L}\right)^{2n}\right]\exp\left[\sum_{n=1}^{\infty}\frac{1}{n!}\xi_n\left(\frac{A_y}{B_0 L}\right)^{n}\right], \tag{3.52}$$

with the coefficients defined by

$$\nu_{2n} = \frac{(-1)^{n+1}2^{2n-1}}{\beta_{pl}}, \quad \xi_n = \frac{2^n}{\beta_{pl}}.$$

We now use the theory of CBPs, as in (Allanson et al. 2015) and Sect. 3.4, to write the pressure as

$$P_{zz} = P_0 \sum_{m=0}^{\infty}\frac{1}{(2m)!}Y_{2m}\left(0, \nu_2, 0, \nu_4, \ldots, 0, \nu_{2m}\right)\left(\frac{A_x}{B_0 L}\right)^{2m}$$
$$\times \sum_{n=0}^{\infty}\frac{1}{n!}Y_n\left(\xi_1, \xi_2, \ldots, \xi_n\right)\left(\frac{A_y}{B_0 L}\right)^{n}.$$

Once again using the simple scaling argument from Eq. (3.28), we have

$$P_{zz} = P_0 \sum_{m=0}^{\infty} \frac{(-1)^m 2^{2m}}{(2m)!} Y_{2m}\left(0 \frac{-1}{2\beta_{pl}}, 0, \frac{-1}{2\beta_{pl}}, \ldots, 0, \frac{-1}{2\beta_{pl}}\right)\left(\frac{A_x}{B_0 L}\right)^{2m}$$
$$\times \sum_{n=0}^{\infty} \frac{2^m}{n!} Y_n\left(\frac{1}{\beta_{pl}}, \frac{1}{\beta_{pl}}, \ldots, \frac{1}{\beta_{pl}}\right)\left(\frac{A_y}{B_0 L}\right)^n.$$

Using the methods established in Chap. 2, namely expansion over Hermite polynomials, we calculate a DF that gives the above pressure

$$f_s = \frac{n_0}{(\sqrt{2\pi}\, v_{\mathrm{th},s})^3} e^{-\beta_s H_s} \times$$
$$\sum_{m=0}^{\infty} a_{2m} \left(\frac{\delta_s}{\sqrt{2}}\right)^{2m} H_{2m}\left(\frac{p_{xs}}{\sqrt{2} m_s v_{\mathrm{th},s}}\right) \times$$
$$\sum_{n=0}^{\infty} b_n \operatorname{sgn}(q_s)^n \left(\frac{\delta_s}{\sqrt{2}}\right)^n H_n\left(\frac{p_{ys}}{\sqrt{2} m_s v_{\mathrm{th},s}}\right), \tag{3.53}$$

for

$$a_{2m} = \frac{(-1)^m 2^{2m}}{(2m)!} Y_{2m}\left(0 \frac{-1}{2\beta_{pl}}, 0, \frac{-1}{2\beta_{pl}}, \ldots, 0, \frac{-1}{2\beta_{pl}}\right),$$
$$b_n = \frac{2^m}{n!} Y_n\left(\frac{1}{\beta_{pl}}, \frac{1}{\beta_{pl}}, \ldots, \frac{1}{\beta_{pl}}\right). \tag{3.54}$$

One can readily calculate the number density for this DF using standard integral results (Gradshteyn and Ryzhik 2007) to be

$$n_s(A_x, A_y) = n_0 \sum_{m=0}^{\infty} a_{2m} \left(\frac{A_x}{B_0 L}\right)^{2m} \sum_{n=0}^{\infty} b_n \left(\frac{A_y}{B_0 L}\right)^n = P_0 \frac{\beta_e \beta_i}{\beta_e + \beta_i}.$$

3.5.3 *Convergence and Boundedness of the DF*

This DF has identical coefficients for the p_{ys}-dependent Hermite polynomials as that derived in Sect. 3.4, and so we need not verify convergence for that series. In fact, all that has changed in the analysis of the coefficients for the p_{xs}-dependent sum is that we now have to consider the Maclaurin coefficients of $\sin^2(A_x/(B_0 L))$ as opposed to $\cos(2A_x/(B_0 L))$. These Maclaurin coefficients both have the same 'factorial dependence' and as such the convergence of the one DF implies the convergence of the other.

The boundedness argument is exactly analogous to that made above for the DF in original gauge, and need not be repeated here.

3.5.4 *Plots of the DF*

We now present plots for the DF given in Eq. (3.53), for $\beta_{pl} = 0.05$ and $\delta_e = \delta_i = 0.03$. This value for β_{pl} is substantially lower than the value used in Sect. 3.4, which had $\beta_{pl} = 0.85$. The ability to go down to lower values of the plasma beta is due to the re-gauging process as explained in Sect. 3.5.1. The plots that we show are intended to demonstrate progress in the numerical evaluation of low-beta DFs for nonlinear force-free fields, and as a proof of principle.

The value of δ_s is chosen such that $\delta_s < \beta_{pl}$, since as explained in Sect. 3.4.4.2, attaining convergence numerically has not been easy for values of $\delta_s > \beta_{pl}$ when $\beta_{pl} < 1$.

Initial investigations of the shape of the variation of the DF in the v_x and v_y directions indicate that the DF seems to have a Gaussian profile, as in the DFs analysed in Sect. 3.4. Hence, as in that work, we shall compare the DFs calculated in this work to drifting Maxwellians, in order to measure the actual difference between the Vlasov equilibrium f_s, and the Maxwellian $f_{Maxw,s}$. In Figs. 3.12a–e and 3.13a–e we give contour plots in $(v_x/v_{\text{th},s}, v_y/v_{\text{th},s})$ space of the 'raw' difference between the DFs defined by Eqs. (3.53) and (3.47). These figures bear close resemblance to those presented in Sect. 3.4.4.2. Specifically, we see 'shallower' peaks for the exact Vlasov solution, f_s, than for $f_{Maxw,s}$. There is also a clear anisotropic effect in that f_s falls off more quickly in the v_x direction than in the v_y direction as compared to $f_{Maxw,s}$. Note that whilst the raw differences plotted in these figures may not seem substantial, they can in fact be significant as a proportion of $f_{Maxw,s}$, and even of the order of the magnitude of $f_{Maxw,s}$. As a demonstration of this fact we present plots in Figs. 3.14a–e and 3.15a–e of the quantity defined by

$$f_{diff,s} = (f_s - f_{Maxw,s})/f_{Maxw,s}$$

for line cuts through $(v_x/v_{\text{th},s}, v_y/v_{\text{th},s} = 0)$ and $(v_x/v_{\text{th},s} = 0, v_y/v_{\text{th},s})$ respectively, for the ions. As suggested by the contour plots, $f_{diff,i}$ takes on significantly larger values in the v_y direction, indicating that the tail of f_i falls off less quickly than $f_{Maxw,i}$ in v_y than in v_x.

We are yet to observe multiple peaks in the multiplicative DFs for the FFHS, derived herein and in Sect. 3.4. However, the summative Harrison-Neukirch equilibria (Harrison and Neukirch 2009a) could develop multiple maxima for sufficiently large values of the magnitude of the drift velocities. For the DF derived in this chapter, and as in Sect. 3.4, the 'amplitude' of the drift velocity profile across the current sheet is given by

$$\frac{u_s}{v_{\text{th},s}} = 2\text{sgn}(q_s)\frac{\delta_s}{\beta_{pl}},$$

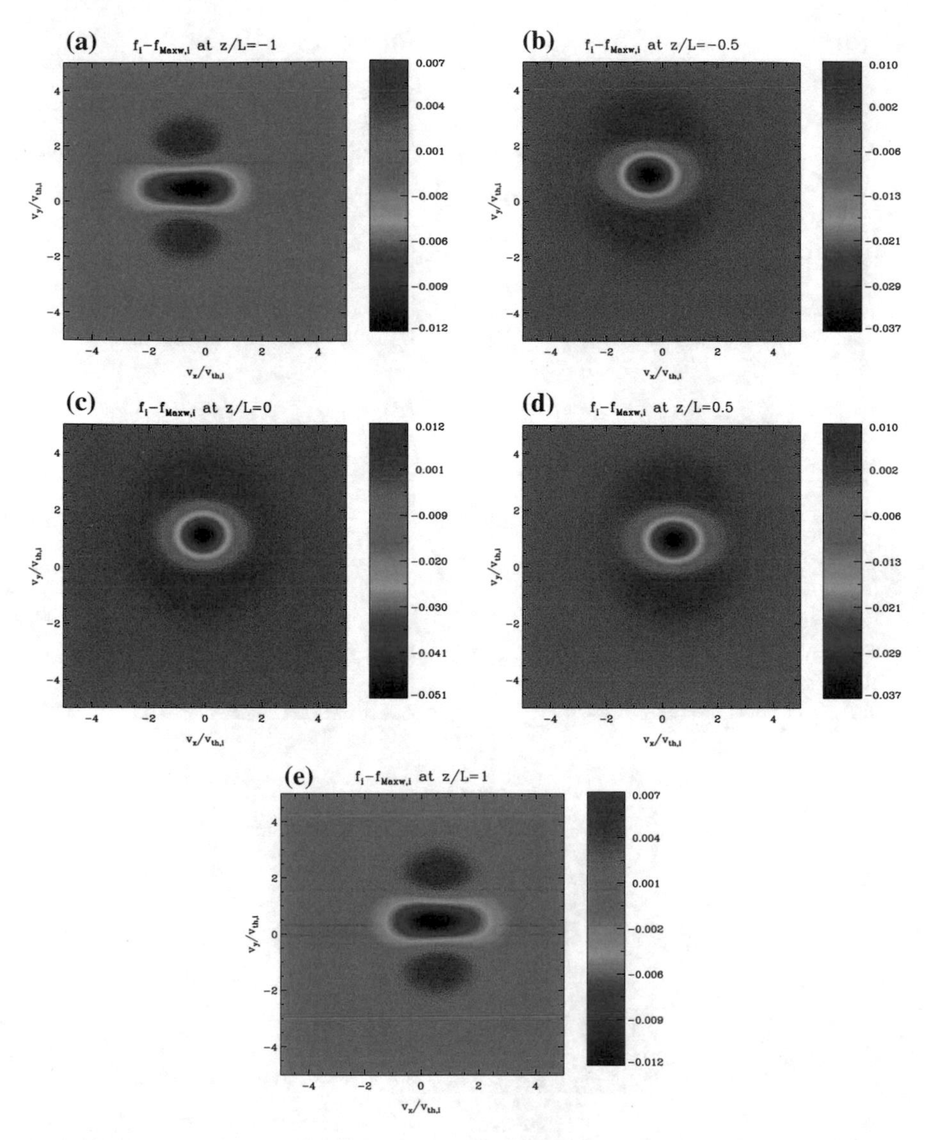

Fig. 3.12 Contour plots of $f_i - f_{Maxw,i}$ for $z/L = -1$ Fig. 3.12a, $z/L = -0.5$ Fig. 3.12b, $z/L = 0$ Fig. 3.12c, $z/L = 0.5$ Fig. 3.12d and $z/L = 1$ Fig. 3.12e. $\beta_{pl} = 0.05$ and $\delta_i = 0.03$

where u_s represents the maximum value of the drift velocities. As a result, large values of the drift velocity correspond to large values of δ_s/β_{pl}, and these are exactly the regimes for which we are struggling to attain numerical convergence. This theory suggests that we may not be seeing DFs with multiple maxima because we are not in the appropriate parameter space.

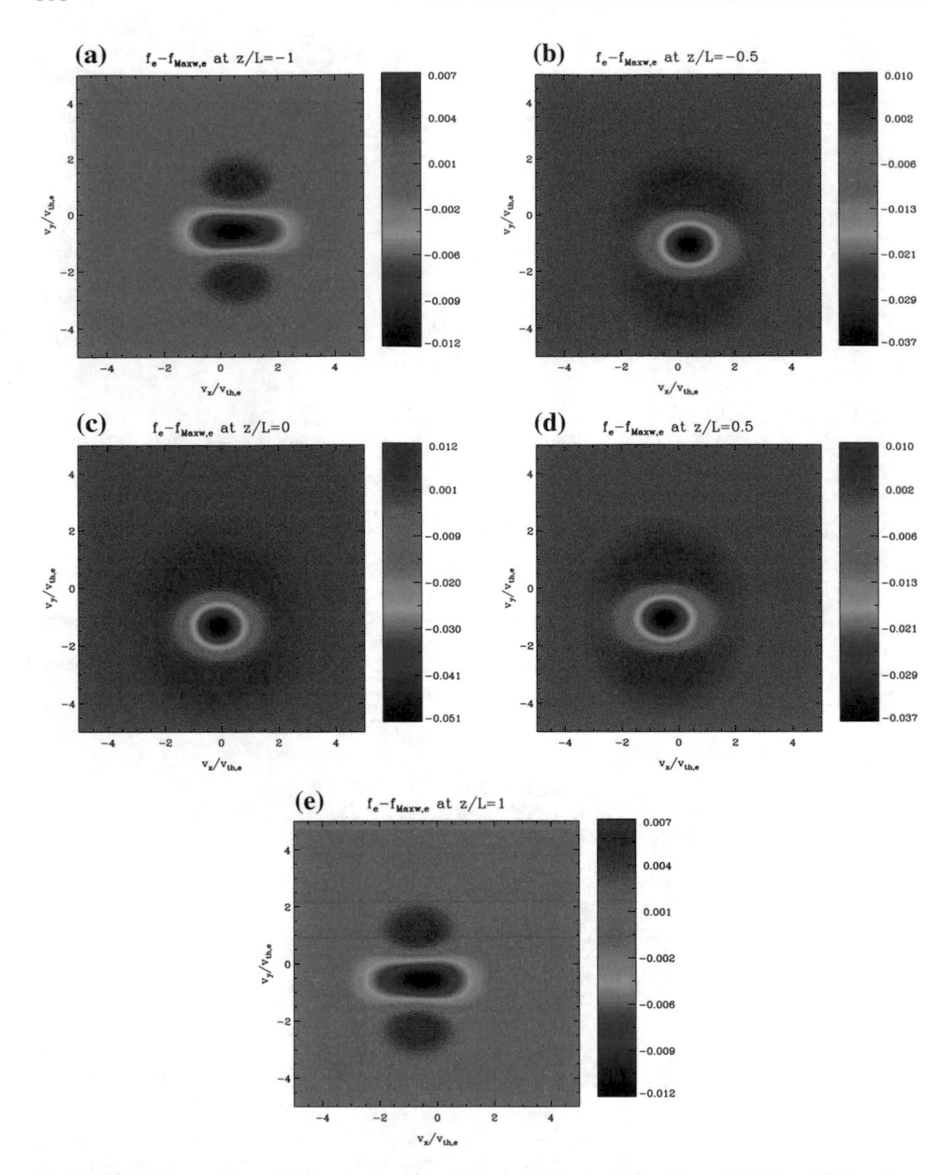

Fig. 3.13 Contour plots of $f_e - f_{Maxw,e}$ for $z/L = -1$ Fig. 3.13a, $z/L = -0.5$ Fig. 3.13b, $z/L = 0$ Fig. 3.13c, $z/L = 0.5$ Fig. 3.13d and $z/L = 1$ Fig. 3.13e. $\beta_{pl} = 0.05$ and $\delta_e = 0.03$

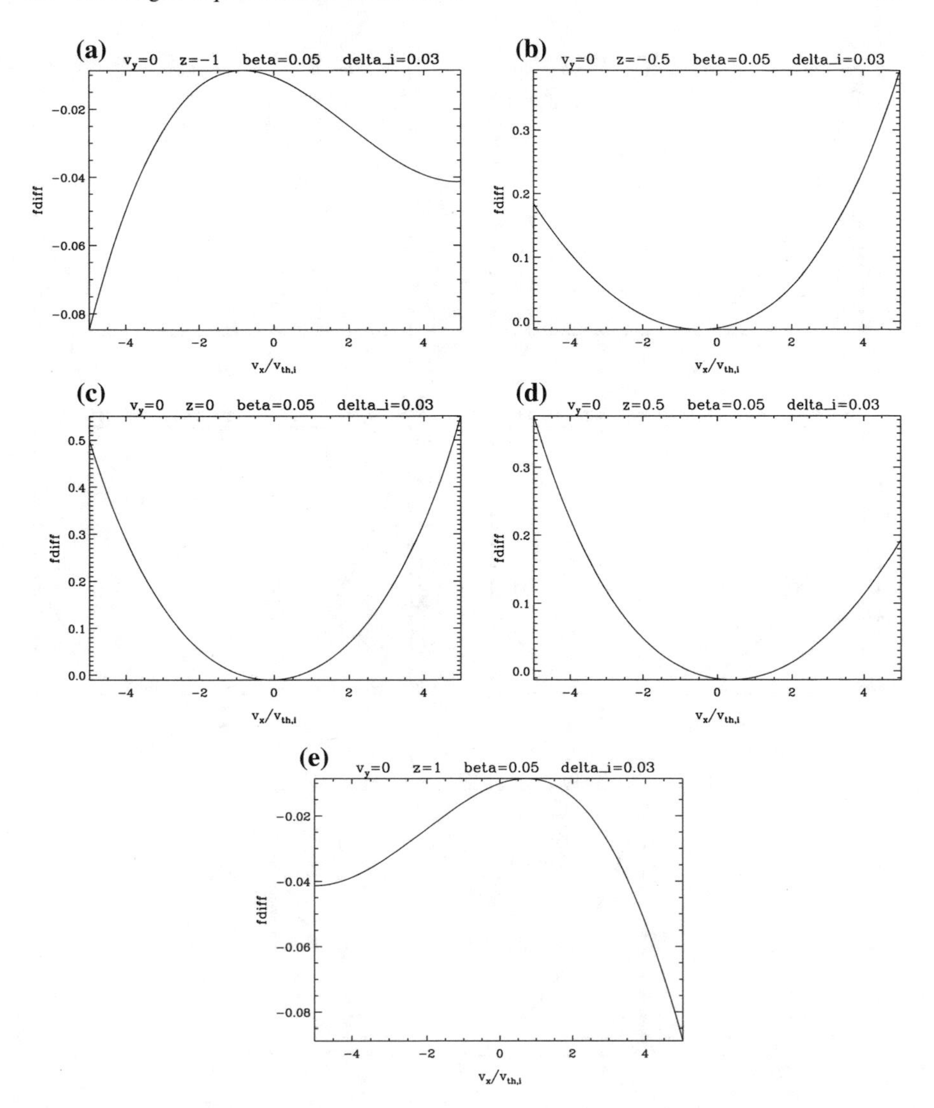

Fig. 3.14 Line plots of $f_{diff,i}$ against $v_x/v_{\text{th},i}$ at $v_y = 0$ for $z/L = -1$ Fig. 3.14a, $z/L = -0.5$ Fig. 3.14b, $z/L = 0$ Fig. 3.14c, $z/L = 0.5$ Fig. 3.14d and $z/L = 1$ Fig. 3.14e. $\beta_{pl} = 0.05$ and $\delta_i = 0.03$

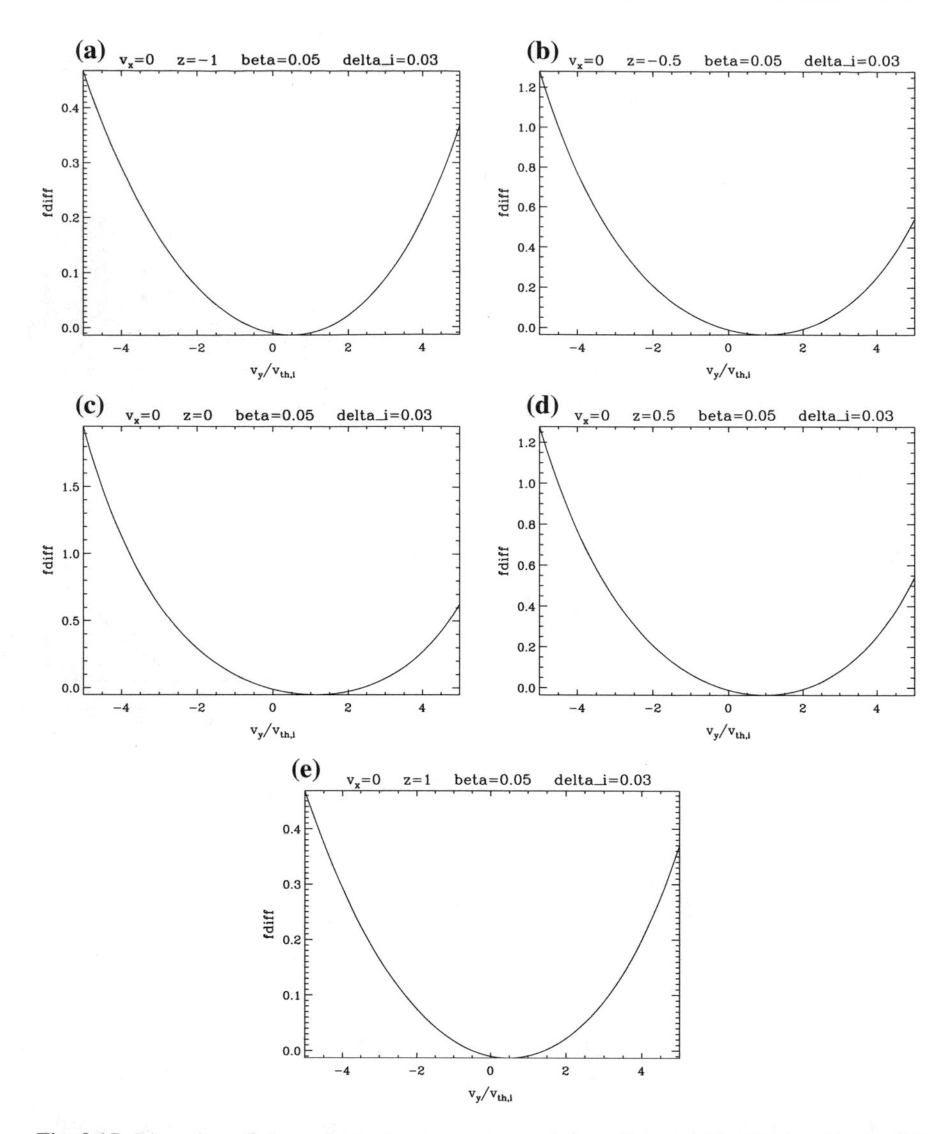

Fig. 3.15 Line plots of $f_{diff,i}$ against $v_y/v_{\text{th},i}$ at $v_x = 0$ for $z/L = -1$ Fig. 3.15a, $z/L = -0.5$ Fig. 3.15b, $z/L = 0$ Fig. 3.15c, $z/L = 0.5$ Fig. 3.15d and $z/L = 1$ Fig. 3.15e. $\beta_{pl} = 0.05$ and $\delta_i = 0.03$

3.6 Summary

This chapter contains presentation and analysis of the first DFs capable of describing low plasma beta, nonlinear force-free collisionless equilibria. By using expressions for the moments of the DFs we have derived the relationships between the micro- and macroscopic parameters of the equilibrium, in particular the current sheet width. We have presented line-plots of the electron DF in the v_x direction as a representative example. These show that the DF has a single maximum in the v_x direction, and *seems* to resemble a Maxwellian, at least for the parameter range studied. However, a detailed comparison with a Maxwellian describing the same particle density and average velocity/current density shows that there are significant deviations. This was corroborated by contour plots of the difference between the DF and the Maxwellian in the (v_x, v_y) plane.

While it has been shown that the infinite series over Hermite polynomials are convergent for all parameter values, plotting the DF in the original gauge,

$$\boldsymbol{A} = B_0 L(2\arctan(\exp(z/L)),\ \ln\operatorname{sech}(z/L),\ 0),$$

has been difficult for the low-beta regime, and particularly due to the v_x dependent sum. As such, $\beta_{pl} = 0.85$ was the lowest value of the plasma beta for which we could be confident in the numerical method. Further work on attaining numerical convergence for a wider parameter range was necessary, with a particular motivation was to find out whether the DF develops multiple peaks similar to the DF found for an additive form of P_{zz} (Neukirch et al. 2009).

Motivated by the numerical challenges mentioned above, in Sect. 3.5 we presented calculations for a DF with a different gauge to that considered in previous studies (Harrison and Neukirch 2009a; Neukirch et al. 2009; Wilson and Neukirch 2011; Abraham-Shrauner 2013; Kolotkov et al. 2015),

$$\boldsymbol{A} = B_0 L(2\arctan\tanh(z/(2L)),\ \ln\operatorname{sech}(z/L),\ 0).$$

We have presented some plots of a comparison between the re-gauged DFs and shifted Maxwellian functions, as a proof of principle, namely that numerical convergence for values of β_{pl} lower than previously reached in the 'original gauge', can now be attained ($\beta_{pl} = 0.05$).

Verification of the analytical properties of convergence and boundedness for both the DFs written as infinite sums over Hermite polynomials have been given. Note that the verification of these DFs is rather involved due to the complex nature of the specific Maclaurin expansions that we consider, and is simpler for more 'straightforward' expansions, e.g. for the example considered in Sect. 2.6.

Future work could involve an in-depth parameter study of the new re-gauged multiplicative DF for the FFHS, with an analysis of how far the exact equilibrium DF differs from an appropriately drifting Maxwellian, frequently used in fully kinetic simulations for reconnection studies. In particular it would be interesting to see how much the DFs differ from drifting Maxwellians as the set of parameters (β_{pl}, δ_s) are varied across a wide range. Preliminary numerical investigations verify that plotting DFs for the FFHS with a lower β_{pl} than previously achieved, namely $\beta_{pl} = 0.05$ rather than $\beta_{pl} = 0.85$, has been made possible by the theoretical developments in this chapter. We have not yet observed multiple maxima for the DFs, but do see significant deviations from Maxwellian distributions, and an anisotropy in velocity space.

References

B. Abraham-Shrauner, Force-free Jacobian equilibria for Vlasov-Maxwell plasmas. Phys. Plasmas **20**(10), 102117 (2013)

O. Allanson, T. Neukirch, S. Troscheit, F. Wilson, From onedimensional fields to Vlasov equilibria: theory and application of Hermite polynomials. J. Plasma Phys. 82.3, 905820306, p. 905820306 (2016)

O. Allanson, T. Neukirch, F.Wilson, S. Troscheit, An exact collisionless equilibrium for the Force-Free Harris Sheet with low plasma beta. Phys. Plasmas 22.10, 102116, p. 102116 (2015)

G. Allen Gary, Plasma beta above a solar active region: rethinking the paradigm. Solar Phys. **203**(1), 71–86 (2001)

A.V. Artemyev, A model of one-dimensional current sheet with parallel currents and normal component of magnetic field. Phys. Plasmas **18**(2), 022104 (2011)

A.V. Artemyev, I.Y. Vasko, S. Kasahara, Thin current sheets in the Jovian magnetotail. Planet. Space Sci. **96**, 133–145 (2014)

N. Attico, F. Pegoraro, Periodic equilibria of the Vlasov-Maxwell system. Phys. Plasmas **6**, 767–770 (1999)

E.T. Bell, Exponential polynomials. Ann. of Math. (2) 35.2, pp. 258–277 (1934)

J. Birn, M. Hesse, Energy release and transfer in guide field reconnection. Phys. Plasmas 17.1, 012109, p. 012109 (2010)

N.A. Bobrova, S.I. Syrovatskii, Violent instability of one-dimensional forceless magnetic field in a rarefied plasma. Soviet J. Exp. Theor. Phys. Lett. **30**, pp. 535-+ (1979)

N.A. Bobrova, S.V. Bulanov, J.I. Sakai, D. Sugiyama, Force-free equilibria and reconnection of the magnetic field lines in collisionless plasma configurations. Phys. Plasmas **8**, 759–768 (2001)

K. Bowers, H. Li., spectral energy transfer and dissipation of magnetic energy from fluid to kinetic scales. Phys. Rev. Lett. 98.3, pp. 035002-+ (2007)

P.J. Channell, Exact Vlasov-Maxwell equilibria with sheared magnetic fields. Phys. Fluids **19**, 1541–1545 (1976)

L. Comtet. Advanced combinatorics: the Art of Finite and Infinite Expansions. In: D. Reidel (ed.) Enlarged Edition, pp. xi+343 (1974)

D.F. Connon, Various applications of the (exponential) complete Bell polynomials. ArXiv e-prints (2010)

D. Correa-Restrepo, D. Pfirsch, Negative-energy waves in an inhomogeneous force-free Vlasov plasma with sheared magnetic field. Phys. Rev. E **47**, 545–563 (1993)

R. Fitzpatrick, Interaction of scrape-off layer currents with magnetohydrodynamical instabilities in tokamak plasmas. Phys. Plasmas 14.6, 062505, p. 062505 (2007)

I.S. Gradshteyn, I.M. Ryzhik, Table of integrals, series, and products. Seventh. (Elsevier/Academic Press, Amsterdam, 2007), pp. xlviii+1171

E.G. Harris, On a plasma sheath separating regions of oppositely directed magnetic field. Nuovo Cimento **23**, 115 (1962)

M.G. Harrison, Equilibrium and dynamics of collisionless current sheets (2009). The University of St Andrews, PhD thesis
M.G. Harrison, T. Neukirch, One-Dimensional Vlasov-Maxwell Equilibrium for the Force-Free Harris Sheet. Phys. Rev. Lett. 102.13, pp. 135003-+ (2009a)
M.G. Harrison, T. Neukirch, Some remarks on one-dimensional forcefree Vlasov-Maxwell equilibria. Phys. Plasmas 16.2, 022106-+ (2009b)
M. Hesse, M. Kuznetsova, K. Schindler, J. Birn, Three-dimensional modeling of electron quasiviscous dissipation in guide-field magnetic reconnection. Phys. Plasmas 12.10, pp. 100704-+ (2005)
W.P. Johnson, The curious history of Faà di Bruno's formula. Amer. Math. Monthly **109**(3), 217–234 (2002)
K.S. Kölbig, The complete Bell polynomials for certain arguments in terms of Stirling numbers of the first kind. J. Comput. Appl. Math. **51**(1), 113–116 (1994)
D.Y. Kolotkov, I.Y. Vasko, V.M. Nakariakov, Kinetic model of forcefree current sheets with non-uniform temperature. Phys. Plasmas 22.11, 112902, p. 112902 (2015)
H. Li, K. Nishimura, D.C. Barnes, S.P. Gary, S.A. Colgate, Magnetic dissipation in a force-free plasma with a sheet-pinch configuration. Phys. Plasmas **10**, 2763–2771 (2003)
G.E. Marsh, *Force-Free Magnetic Fields: Solutions* (Topology and Applications. World Scientific, Singapore, 1996)
E. Moratz, E.W. Richter, Elektronen-Geschwindigkeitsverteilungsfunktionen für kraftfreie bzw. teilweise kraftfreie Magnetfelder. Zeitschrift Naturforschung Teil A 21, p. 1963 (1966)
P.M. Morse, H. Feshbach, Methods of theoretical physics. 2 volumes. (McGraw-Hill Book Co. 1953)
H.E. Mynick, W.M. Sharp, A.N. Kaufman, Realistic Vlasov slab equilibria with magnetic shear. Phys. Fluids **22**, 1478–1484 (1979)
T. Neukirch, Magnetic field extrapolation. In: D.E. Innes, A. Lagg, S.A. Solanki (eds.) Chromospheric and Coronal Magnetic Fields. vol. 596. ESA Special Publication (2005)
T. Neukirch, F. Wilson, M.G. Harrison, A detailed investigation of the properties of a Vlasov-Maxwell equilibrium for the force-free Harris sheet. Phys. Plasmas **16**(12), 122102 (2009)
K. Nishimura, S.P. Gary, H. Li, S.A. Colgate, Magnetic reconnection in a force-free plasma: simulations of micro- and macroinstabilities. Phys. Plasmas **10**, 347–356 (2003)
E.V. Panov, A.V. Artemyev, R. Nakamura, W. Baumjohann, Two types of tangential magnetopause current sheets: cluster observations and theory. J. Geophys. Res. (Space Physics) 116, A12204, A12204 (2011)
A. Petrukovich, A. Artemyev, I. Vasko, R. Nakamura, L. Zelenyi, Current Sheets in the Earth Magnetotail: Plasma and magnetic field structure with cluster project observations. Space Sci. Rev. **188**, 311–337 (2015)
E. Priest, T. Forbes, *Magnetic Reconnection* (Cambridge University Press, Cambridge, UK, 2000)
P.L. Pritchett, F.V. Coroniti, Three-dimensional collisionless magnetic reconnection in the presence of a guide field. J. Geophys. Res. (Space Physics) 109.A18, pp. 1220-+ (2004)
J. Riordan, An introduction to combinatorial analysis. Wiley Publications in Mathematical Statistics. John Wiley & Sons, Inc., New York; Chapman & Hall, Ltd., London, pp. xi+244 (1958)
J.-I. Sakai, A. Matsuo, Three-dimensional dynamics of relativistic flows in pair plasmas with force-free magnetic configuration. Phys. Plasmas **11**(6), 3251–3258 (2004)
T. Sakurai, Computational modeling of magnetic fields in solar active regions. Space Sci. Rev. **51**, 11–48 (1989)
K. Schindler, *Physics of Space Plasma Activity*. (Cambridge University Press, 2007)
A. Sestero, Self-consistent description of a warm stationary plasma in a uniformly sheared magnetic field. Phys. Fluids **10**, 193–197 (1967)
J.B. Taylor, Relaxation and magnetic reconnection in plasmas. Rev. Mod. Phys. **58**, 741–763 (1986)
J.B. Taylor, Relaxation of Toroidal Plasma and generation of reverse magnetic fields. Phys. Rev. Lett. **33**, 1139–1141 (1974)

I.Y. Vasko, A.V. Artemyev, A.A. Petrukovich, H.V. Malova, Thin current sheets with strong bell-shape guide field: cluster observations and models with beams. Ann. Geophys. **32**, 1349–1360 (2014)
G.E. Vekstein, N.A. Bobrova, S.V. Bulanov, On the motion of charged particles in a sheared force-free magnetic field. J. Plasma Phys. **67**, 215–221 (2002)
T. Wiegelmann, T. Sakurai, solar force-free magnetic fields. Living Rev. Solar Phys. 9.5 (2012)
F. Wilson, Equilibrium and stability properties of collisionless current sheet models (2013). The University of St Andrews, PhD thesis
F. Wilson, O. Allanson, T. Neukirch, The collisionless tearing mode in a force-free current sheet (In preparation) (2017)
F. Wilson, T. Neukirch, M. Hesse, M.G. Harrison, C.R. Stark, Particlein- cell simulations of collisionless magnetic reconnection with a nonuniform guide field. Phys. Plasmas 23.3, p. 032302 (2016)
F. Wilson, T. Neukirch, A family of one-dimensional Vlasov-Maxwell equilibria for the force-free Harris sheet. Phys. Plasmas **18**(8), 082108 (2011)

Chapter 4
One-Dimensional Asymmetric Current Sheets

Reconnection is now among the most fundamental unifying concepts in astrophysics, comparable in scope and importance to the role of natural selection in biology.

from Moore, Burch, and Torbert, (2015)

Much of the work in this chapter is drawn from Allanson et al. (2017)

4.1 Preamble

The NASA MMS mission has very recently made in situ diffusion region measurements of asymmetric magnetic reconnection for the first time (Burch et al. 2016). In order to compare to the data obtained from kinetic-scale observations (e.g. see Burch and Phan 2016), it would be useful to have initial equilibrium conditions for PIC simulations that reproduce the physics of the dayside magnetopause current sheet as accurately as possible, i.e. self-consistent VM equilibria that model the magnetosheath-magnetosphere asymmetries in pressure and magnetic field strength.

In this chapter, we present new 'exact numerical' (numerical solutions to equations for exact VM equilibria), and exact analytical equilibrium solutions of the VM system that are self-consistent with 1D and asymmetric Harris-type current sheets, with a constant guide field. The DFs can be represented as a combination of shifted Maxwellian DFs, are consistent with a magnetic field configuration with more freedom than the previously known exact solution (Alpers 1969), and have different bulk flow properties far from the sheet.

O. Allanson, *Theory of One-Dimensional Vlasov-Maxwell Equilibria*, Springer Theses, https://doi.org/10.1007/978-3-319-97541-2_4

4.2 Introduction

4.2.1 *Asymmetric Current Sheets*

Under many circumstances (and unlike the application in Chap. 3), the plasma conditions can be different on either side of the current sheet, e.g. the magnetic field strength and its orientation. As well as in the magnetopause (e.g. see Burch and Phan 2016), such asymmetric current sheets are observed at Earth's magnetotail (e.g. Øieroset et al. 2004), in the solar wind (e.g. Gosling et al. 2006), between solar flux tubes (e.g. Linton 2006; Murphy et al. 2012; Zhu et al. 2015), in turbulent plasmas (e.g. Servidio et al. 2009; Karimabadi et al. 2013), and inside a tokamak (e.g. Kadomtsev 1975).

Regarding the theoretical modelling of dynamical features, various authors have considered the impact of asymmetric current sheets on different aspects of instability and magnetic reconnection, such as the 'Sweet-Parker' style analysis carried out by Cassak and Shay (2007); the development of current driven instabilities (the lower-hybrid instability) (Roytershteyn et al. 2012); and the suppression of reconnection at Earth's magnetopause (Swisdak et al. 2003; Phan et al. 2013; Trenchi et al. 2015; Liu and Hesse 2016). Whilst it can be argued that the general properties (e.g. the reconnection rate) of the nonlinear phase physics of magnetic reconnection are relatively insensitive with regards to the exactitude of the initial conditions, the physics in the linear stage can affect the dynamical evolution of the current sheets, and that can only be confidently studied with exact initial conditions (e.g. see Dargent 2016).

To give some specific examples of the use of exact solutions, setting up a VM equilibrium current sheet in numerical simulations would be helpful for the study of collisionless tearing instabilities, which could be important to understand the role of tearing modes in determining the orientation of the three-dimensional reconnection x-line in an asymmetric geometry (Liu et al. 2015). This is especially crucial for predicting the location of magnetic reconnection at Earth's magnetopause under diverse solar wind conditions, as discussed in (Komar et al. 2015) for example. Knowledge of an exact equilibirum also facilitates the study of tearing instabilities under the influence of cross-sheet gradients (e.g. see Zakharov and Rogers 1992; Kobayashi et al. 2014; Pueschel et al. 2015; Liu and Hesse 2016), which can be important for understanding the onset and diamagnetic suppression of sawtooth crashes in fusion devices.

4.2.2 *Modelling the Magnetopause Current Sheet*

4.2.2.1 Model Paradigm

The macroscopic equilibrium for which we wish to obtain a self-consistent VM equilibrium is that which describes a current sheet in the Earth's dayside magnetopause. Figure 4.1 depicts the Earth's magnetopause, its relation to the rest of the Earth's

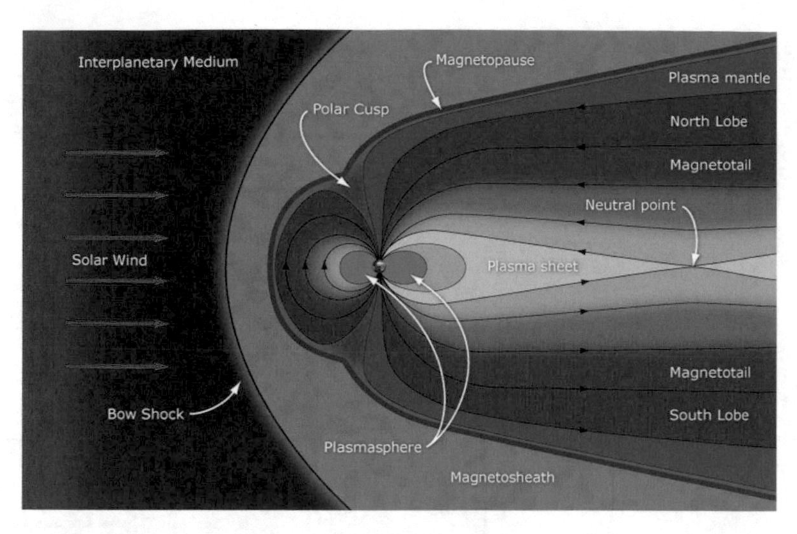

Fig. 4.1 Diagram representing the Earth's magnetic environment, and it's interaction with the solar wind. Image credit: NASA, and without copyright

magnetosphere, and the interaction with the solar wind. In line with other theoretical approaches (e.g. see Hesse et al. 2013) and observational (e.g. see Burch et al. 2016) conclusions, the equilibrium should be 'asymmetric' with respect to either side of the current sheet, i.e. it should be characterised by an enhanced density/pressure on the magnetosheath side of the current sheet, and an enhanced magnetic field magnitude on the magnetosphere side. These basic requirements are shown by Fig. 4.2a and b, in which the coordinates (x, y, z) are related to the "Boundary Normal" coordinates, LMN, (e.g. see Hapgood 1992; Burch et al. 2016). Their correspondence is given by $(\hat{x}, \hat{y}, \hat{z}) \sim (\hat{L}, \hat{M}, \hat{N})$, with the xy plane tangential to the magnetopause, and z normal to it. As explained by Hapgood (1992), "*There is no universal convention to resolve the L and M axes. The relationship between LMN and other systems ... is dependent on position.*" For a heuristic understanding, and in the paradigm of the 'square-on' geometry presented by Fig. 4.1, we can think of $x \sim L$ as pointing 'Earth North', $y \sim M$ as pointing 'Earth West', and $z \sim N$ as pointing 'Sunward'. The figures relate to a specific magnetic field, to be defined in Sect. 4.3.1.2, but they portray the basic features that the model should have. Essentially, pressure balance dictates that an enhanced magnitude of magnetic pressure on the magnetosphere side of the current sheet ($z < 0$, $N < 0$) relies on a depleted thermal pressure, and vice versa for the magnetosheath side ($z > 0$, $N > 0$). However, the current density is modelled to be symmetric. As in Chaps. 2 and 3, we assume a 1D geometry for which $\nabla = (0, 0, \partial/\partial z)$, which is justifiable by a separation of scales (e.g. see Quest and Coroniti 1981). In this case, a quasineutral macroscopic equilibrium will obey the following equation,

$$\frac{d}{dz}\left(P_{zz} + \frac{B^2}{2\mu_0}\right) = 0, \tag{4.1}$$

Fig. 4.2 The AH+G equilibrium configuration (Eq. 4.7)

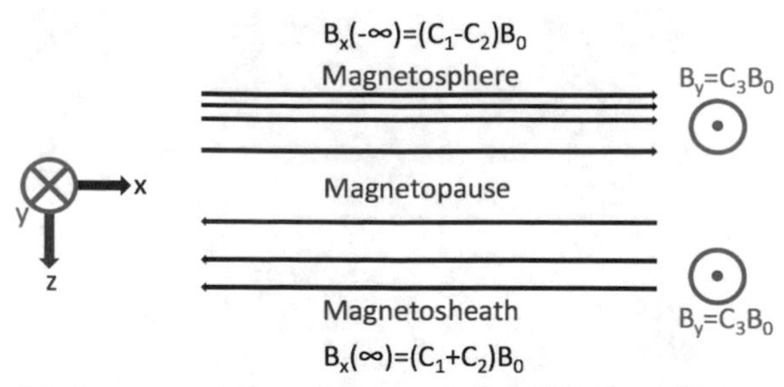

(a) A representative diagram of the AH+G equilibrium magnetic field, for $C_1 + C_2 < C_1 - C_2$.

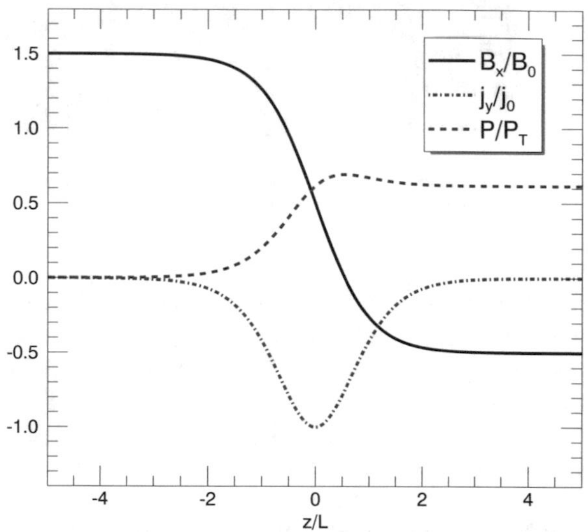

(b) Normalised magnetic field $\tilde{B}_x$, current density $\tilde{j}_y$, and scalar pressure $\tilde{p}$ for parameter values $C_1 = 0.5$, $C_2 = -1$, $C_3 = 1$ and $P_T = (C_1^2 + C_2^2 + C_3^2 - 2C_1C_2)/2 = 1.625$.

but, in contrast to the application to force-free current sheets in Chap. 3, P_{zz} and B^2 must be non-uniform in z.

4.2.2.2 Typical Approach in PIC Simulations

In the effort to model dayside magnetopause reconnection, asymmetric macroscopic equilibria that satisfy Eq. (4.1) have been used in PIC simulations by e.g. Swisdak et al. (2003), Pritchett (2008), Huang et al. (2008), Malakit et al. (2010), Wang et al. (2013), Aunai et al. (2013b), Aunai et al. (2013a), Hesse et al. (2013), Hesse et al. (2014), Dargent (2016), Liu and Hesse (2016). All but two (Aunai et al. 2013a; Dargent 2016) of these studies have used drifting Maxwellian DFs as initial conditions (Eq. 1.26). As discussed in more detail in Sect. 1.3.3.1, these DFs can reproduce the same moments ($n(z)$, $\boldsymbol{V}_s(z)$, $p(z)$) necessary for a quasineutral fluid equilibrium, but are not exact solutions of the Vlasov equation and hence do not describe a kinetic equilibrium. The main aim of this chapter is to calculate exact solutions of the equilibrium VM equations consistent with a suitable dayside magnetopause current sheet model, in order to circumvent the need to use non-equilibrium DFs of the form in Eq. (1.26).

The work in this chapter is relevant to the main focus of the MMS mission, i.e. asymmetric magnetic reconnection, and so we envisage that this could be the main use of the results at the present time. However, as mentioned in Sect. 4.2.1, there are other potential applications of the work to basic equilibrium and instability physics in the magnetotail, solar corona, turbulent plasmas and tokamaks.

4.3 Exact VM Equilibria for 1D Asymmetric Current Sheets

4.3.1 Theoretical Obstacles

The (symmetric) Harris sheet (Eq. 1.24) can be rendered asymmetric—the *asymmetric Harris sheet* (AHS)—by the simple addition of a constant component to B_x,

$$\boldsymbol{B} = B_0\left(C_1 + C_2\tanh\left(\frac{z}{L}\right), 0, 0\right), \tag{4.2}$$

for C_1 and C_2 dimensionless constants, and and for which there is a field reversal (a change in the sign of B_x) only when

$$\left|\frac{C_1}{C_2}\right| < 1. \tag{4.3}$$

The current density, j_y, is indepdendent of C_1, and so whilst a field-reversal is not essential for the existence of a current sheet in itself, we shall only consider the field-reversal regime. The addition of C_1 to B_x leads to an equilibrium described by

$$\begin{aligned} \frac{B^2}{2\mu_0}(z) &= \frac{B_0^2}{2\mu_0}\left(C_1^2 + 2C_1C_2\tanh\left(\frac{z}{L}\right) + C_2^2\tanh^2\left(\frac{z}{L}\right)\right), \\ P_{zz}(z) &= P_T - \frac{B_0^2}{2\mu_0}\left(C_1^2 + 2C_1C_2\tanh\left(\frac{z}{L}\right) + C_2^2\tanh^2\left(\frac{z}{L}\right)\right), \end{aligned} \tag{4.4}$$

with $P_T > B_0^2(|C_1| + |C_2|)^2/(2\mu_0)$ the constant total pressure.

The VM equilibrium DF self-consistent with the Harris sheet (Harris 1962 and as discussed in Sect. 1.3.2.1),

$$f_s = \frac{n_{0s}}{(\sqrt{2\pi}\, v_{th,s})^3} e^{-\beta_s (H_s - u_{ys} p_{ys})},$$

can also be made to be consistent with the field,

$$\boldsymbol{B} = B_0 \left(\tanh\left(\frac{z}{L}\right), C_3, 0 \right),$$

i.e. a Harris sheet plus guide field. This is achieved fairly simply by 'sending' $A_x = 0 \to A_x = C_3 B_0 z$. This adds no real complications since $j_x = 0$, $P_{zz}(A_y)$ remains unchanged, and one essentially just solves Ampère's Law with different conditions as $|z| \to \infty$,

$$\nabla^2 A_x = 0 \;\; \text{s.t.} \;\; A_x = 0 \;\; \to \nabla^2 A_x = 0 \;\; \text{s.t.} \;\; A_x = C_3 B_0 z.$$

In the analogy of the particle in a potential (see Sect. 1.3.5.3), this corresponds to the particle having a non-zero and constant component of 'velocity' in the $x \sim A_x$ direction, instead of zero velocity in that direction. As a result, one might expect that it should be relatively straightforward to adapt the Harris DF to be self-consistent with the AHS, but this is not the case in the field-reversal regime.

4.3.1.1 P_{zz} Must Depend on Both A_x and A_y

The AHS has only one component of the current density, j_y, and since $\boldsymbol{j} = \partial P_{zz}/\partial \boldsymbol{A}$, one might expect that the equilibrium could be described by $P_{zz} = P_{zz}(A_y)$, and hence $f_s = f_s(H_s, p_{ys})$ accordingly. However, using the analogy of the particle in a potential (see Sect. 1.3.5.3), in which the following correspondences hold

$$\begin{aligned} \text{Position: } (x, y) &\sim (A_x, A_y), \\ \text{Time: } t &\sim z \\ \text{Velocity: } (v_x(t), v_y(t)) &\sim \left(\frac{dA_x}{dz}(z), \frac{dA_y}{dz}(z) \right) \sim (B_y, -B_x), \\ \text{Potential: } \mathcal{V}(x, y) &\sim P_{zz}(A_x, A_y), \\ \text{Force: } \boldsymbol{\mathcal{F}}(x(t), y(t)) &\sim \mu_0 \frac{d^2 \boldsymbol{A}}{dz^2}, \\ \text{Equation of motion: } \boldsymbol{\mathcal{F}} = -\nabla \mathcal{V} &\sim \mu_0 \frac{d^2 \boldsymbol{A}}{dz^2} = -\frac{\partial P_{zz}}{\partial \boldsymbol{A}}, \end{aligned}$$

we note that—crucially—velocity is conjugate to the derivatives of A_x, A_y, and hence the magnetic field. The important observation to make is that a single-valued

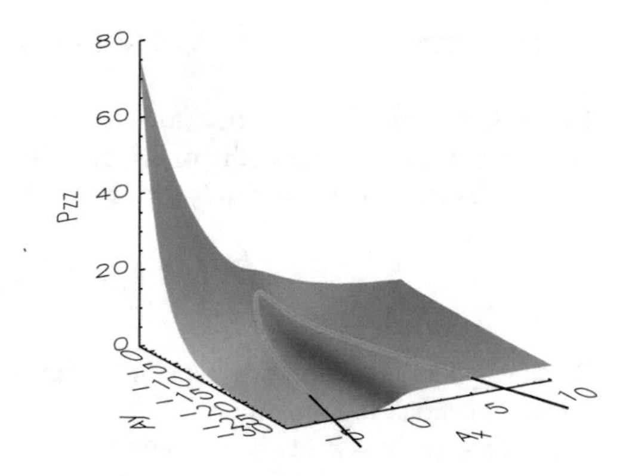

Fig. 4.3 The "tanh" pressure function for $C_1 = -1, C_2 = 8, C_3 = 1, C_2 = 2\sqrt{C_1^2} = 2$

and 1D potential, $P_{zz}(A_y)$, cannot be compatible with a 'velocity' of the form of the magnetic field in Eq. (4.2),

$$v_y(t) \sim C_1 + C_2 \tanh t,$$

when we are in the field-reversal regime ($|C_1| < |C_2|$). The reasoning is as follows.

Without loss of generality suppose that $C_1, C_2 > 0$. The particle begins its journey at $t = -\infty$, $y = \infty$ with velocity $C_1 - C_2 < 0$. It then rolls up a 'hill' in the potential, is stationary at $t = \tanh^{-1}(-C_1/C_2)$, and rolls back down the hill towards $y = \infty$ with final velocity $C_1 + C_2$ at $t = \infty$. This trajectory is not possible for a conservative potential that is single-valued in space. Hence we conclude that a 1D asymmetric current sheet with field reversal can not be analytically self-consistent with a pressure tensor that is a function of only one component of the vector potential.

Despite the fact that $j_x = 0$ for the AHS, and hence $\partial P_{zz}/\partial A_x = 0$, it has become apparent that we require the 'hill' to be 2D, such as the $P_{zz}(A_x, A_y)$ function depicted in Fig. 4.3, for which the overlaid line depicts the particle trajectory. (The exact form and derivation of that particular pressure function shall be discussed in Sect. 4.3.2).

We note that 'exact numerical' VM equilibria have recently been found by Belmont et al. (2012), Dorville et al. (2015), using the inverse approach, for the 'normal/symmetric' Harris sheet magnetic field, and a modified 'force-free Harris sheet' respectively. The equilibria have asymmetries in the number density and temperature either side of the sheet, with Dorville et al. (2015) including an electric field. Their methods rely on similar notions to those discussed above, for which the DFs were multi-valued functions of the constants of motion. The DF derived by Belmont et al. (2012) has been used as the initial condition for Hybrid simulations by Aunai et al. (2013a), and PIC simulations by Dargent (2016). Exact numerical solutions for asymmetric current sheets are more numerous for the forward problem, with examples in e.g. Kan (1972), Lemaire and Burlaga (1976), Kuznetsova and Roth (1995), Roth et al. (1996), Lee and Kan (1979).

4.3.1.2 Prior Exact Analytical VM Equilibria

To our knowledge, there is one known exact VM equilibrium for a magnetic field like the AHS. In the Appendix of Alpers (1969), a DF is derived that is consistent with the 'Alpers magnetic field', which could be written in a z-dependent geometry as

$$\boldsymbol{B} = B_0\left(-\frac{B_2}{2}\left(1+\tanh\left(\frac{z}{L}\right)\right), \frac{B_1}{2}\tanh\left(\frac{z}{L}\right), 0\right). \tag{4.5}$$

Despite appearances, this magnetic field is almost equivalent to the AHS. To see this, we make a small digression.

First allow the AHS to have a constant *guide field*, with $\boldsymbol{A}$, $\boldsymbol{B}$ and $\boldsymbol{j}$ given by

$$\boldsymbol{A} = B_0 L(C_3\tilde{z},\ -C_1\tilde{z} - C_2\ln\cosh\tilde{z},\ 0), \tag{4.6}$$

$$\nabla\times\boldsymbol{A} = \boldsymbol{B} = B_0(C_1 + C_2\tanh\tilde{z},\ C_3,\ 0), \tag{4.7}$$

$$\frac{1}{\mu_0}\nabla\times\boldsymbol{B} = \boldsymbol{j} = \frac{B_0}{\mu_0 L}(0,\ C_2\text{sech}^2\tilde{z},\ 0), \tag{4.8}$$

then we have the *Asymmetric Harris sheet plus guide field* (AH+G), with C_3 a non-zero constant. The vector potential, magnetic field, current density and length scales are normalised according to $\tilde{\boldsymbol{A}}B_0L = \boldsymbol{A}$, $\tilde{\boldsymbol{B}}B_0 = \boldsymbol{B}$, $\boldsymbol{j} = j_0\tilde{\boldsymbol{j}}$ and $z = L\tilde{z}$ respectively, with $j_0 = B_0/(\mu_0 L)$. Example profiles of $\tilde{B}_x$ and $\tilde{j}_y$ are plotted in Fig. 4.2b for parameter values $C_1 = 0.5$, $C_2 = -1$ and $C_3 = 1$, (in line with other theoretical studies, e.g. see Pritchett 2008; Liu and Hesse 2016). For these parameter values, the left and right hand sides of the plot represent the magnetosphere and magnetosheath respectively, whilst the central current sheet is in the magnetopause. The equilibrium is maintained by the 'gradient of a scalar pressure', $p(z) := P_{zz}$, according to

$$P_{zz}(\tilde{z}) = P_T - \frac{B_0^2}{2\mu_0}\left(C_1^2 + 2C_1C_2\tanh\tilde{z} + C_2^2\tanh^2\tilde{z} + C_3^2\right), \tag{4.9}$$

for P_T the total pressure (magnetic plus thermal), and $P_{zz} > 0$ for $C_1^2 + 2|C_1C_2| + C_2^2 + C_3^2 < 2\mu_0 P_T/B_0^2$. The profile of $\tilde{p}(\tilde{z}) = P_{zz}/P_T$ is plotted in Fig. 4.2b, for $P_T = 1.625$.

After a rotation by $\tan\theta = C_1/C_3$, the AH+G field becomes

$$\mathbf{B}' = B_0\left(\frac{C_2C_3}{\sqrt{C_1^2+C_3^2}}\tanh\tilde{z}, \sqrt{C_1^2+C_3^2} + \frac{C_1C_2}{\sqrt{C_1^2+C_3^2}}\tanh\tilde{z}, 0\right), \tag{4.10}$$

which is essentially equivalent to the Alpers magnetic field in Eq. (4.5) when $C_1C_2 = C_1^2 + C_3^2$. As such, the Alpers magnetic field is very similar to the AH+G field, but with one fewer degree of freedom.

For the DF derived by Alpers, and those to be developed in this chapter, the guide field, $B_y = C_3 B_0$, is crucial for making analytical progress. The existence of B_y necessitates a non-trivial $A_x = C_3 B_0 z$, and as a result the 'potential' P_{zz} can now be a function of both A_x and A_y. This two-dimensionality was reasoned to be an important feature of analytically described asymmetric fields in Sect. 4.3.1.1, and will allow us to construct exact analytical DFs.

There is one more difference between the equilibrium derived by Alpers, and the one that we shall consider, and it is related to the bulk flows.

As is necessary for consistency between the microscopic and macroscopic descriptions, the Alpers DF is self-consistent with the prescribed magnetic field, i.e. the sum of the individual species (kinetic) currents are equal to the current prescribed by Ampère's Law, i.e. $\sum_s \boldsymbol{j}_s = \boldsymbol{j} = \nabla \times \boldsymbol{B}/\mu_0$. However, the $\boldsymbol{j}_s$ are non-zero at $z = +\infty$ (in our co-ordinates), i.e. the magnetosheath side. In contrast, Eq. (5.19) shows that the macroscopic current densities vanish as $z \to \pm\infty$, i.e. the Alpers DF gives species currents $\boldsymbol{j}_s$ that are not proportional to the macroscopic current $\boldsymbol{j}$. That is to say that there is finite ion and electron mass flow at infinity, "*impinging vertically*" on the magnetosheath side of the current sheet. This could be appropriate if one wishes to consider a larger scale/global model including bulk flows at the boundary, but it is not suitable if one wishes to consider the domain as an isolated 'patch', representing a local current sheet structure.

In summary, the DF that we derive shall be consistent macroscopically with an equilibrium for which there are no mass flows at the boundary (as typically assumed in PIC simulations, e.g. Aunai et al. 2013b; Hesse et al. 2013), and is self-consistent with a magnetic field that has more degrees of freedom than that in Alpers (1969).

4.3.2 Outline of Basic Method

In order to find a VM equilibrium, we shall use 'Channell's method' (Channell 1976). As discussed in Chap. 1, this involves the following steps:

Pressure tensor: First calculate a functional form $P_{zz}(A_x, A_y)$ that 'reproduces' the scalar pressure of Eq. (4.9) as a function of z. It must also satisfy $\partial P_{zz}/\partial \boldsymbol{A} = \boldsymbol{j}(z)$. There could in principle be infinitely many functions $P_{zz}(A_x, A_y)$ that satisfy both these criteria, but we shall choose specific $P_{zz}(A_x, A_y)$ functions which allows us to make analytical progress.

Note that this procedure is—by the analogy of a particle in a potential—contrary to the 'typical approach', in which one tries to establish the trajectory in a given potential. We know the 'trajectory as a function of time' $\boldsymbol{A}(z)$, and the value of the potential along it $P_{zz}(z)$, and seek to construct a self-consistent 'potential function in space', $P_{zz}(A_x, A_y)$.

Inversion: The second step is to use the assumed form of the DF in Eq. (1.37) in the definition of the pressure tensor component P_{zz} as the second-order velocity moment of the DF, $P_{zz} = \sum_s m_s \int v_z^2 f_s d^3v$, and attempt to invert the integral transforms, either by Fourier transforms, Hermite polynomials, or perhaps some other method.

Macro-micro: The inversion process must yield an f_s that not only reproduces the macroscopic expression for the pressure tensor (achieved by fixing parameters), but also that is consistent with quasineutrality ($\sigma(A_x, A_y) = 0$), and in this case strict neutrality, $\phi = 0$.

Let us first consider possible expressions for $P_{zz}(A_x, A_y)$. Pressure balance dictates that

$$P_{zz}(\tilde{z}) = P_T - \frac{B_0^2}{2\mu_0}\left(C_1^2 + 2C_1C_2 \tanh\tilde{z} + C_2^2 \tanh^2\tilde{z} + C_3^2\right). \tag{4.11}$$

Using the knowledge that exponential functions are eigenfunctions of the Weierstrass transform (Wolf 1977), we would like to use exponential functions to represent the P_{zz} function wherever possible. In Sects. 4.4 and 4.5 we present two different attempts at using Channell's method for the AH+G field. The first requires a numerical approach, whereas the second can be completed analytically.

4.4 The Numerical/"tanh" Equilibrium DF

4.4.1 The Pressure Function

From Eq. (4.6) we see that $\exp(2A_y/(C_2B_0L)) = \text{sech}^2\tilde{z}\exp(-2C_1\tilde{z}/C_2)$, and so we can construct one part of the RHS of Eq. (4.11) by

$$\tanh^2\tilde{z} = 1 - \text{sech}^2\tilde{z} = 1 - \exp\left(\frac{2\tilde{A}_y}{C_2}\right)\exp\left(\frac{2C_1\tilde{A}_x}{C_2C_3}\right). \tag{4.12}$$

The remaining task is to invert $\tanh\tilde{z} = \tanh\tilde{z}(\tilde{A}_x, \tilde{A}_y)$, and this is most readily achieved by

$$\tanh\tilde{z} = \tanh\left(\frac{\tilde{A}_x}{C_3}\right). \tag{4.13}$$

Note that we have not chosen to take the square root of Eq. (4.12), since we—naively—expect to be able to invert the Weierstrass transform for the expression in Eq. (4.13) more easily (and in fact, it can be shown that one cannot solve Ampère's Law by doing so). Substituting Eqs. (4.13) and (4.12) into Eq. (4.11) gives the pressure tensor

$$P_{zz} = P_0 \left[C_2 \exp\left(\frac{2\tilde{A}_y}{C_2}\right) \exp\left(\frac{2C_1\tilde{A}_x}{C_2C_3}\right) - 2C_1 \tanh\left(\frac{\tilde{A}_x}{C_3}\right) + C_b \right], \qquad (4.14)$$

with $C_b > 2C_1$ for positivity of the pressure. There is *a priori* no guarantee that this pressure tensor will satisfy Ampère's law, $\partial P_{zz}/\partial \boldsymbol{A} = \boldsymbol{j}$. We can check the validity of the pressure with respect to Ampère's law, by

$$\frac{\partial P_{zz}}{\partial A_x} = \frac{P_0}{B_0 L}\frac{\partial \tilde{P}_{zz}}{\partial \tilde{A}_x} = 2\frac{P_0}{B_0 L}\frac{C_1}{C_3}\left(e^{2\tilde{A}_y/C_2}e^{2C_1\tilde{A}_x/(C_2C_3)} - \operatorname{sech}^2(\tilde{A}_x/C_3)\right) = 0 = j_x,$$

and

$$\frac{\partial P_{zz}}{\partial A_y} = \frac{P_0}{B_0 L}\frac{\partial \tilde{P}_{zz}}{\partial \tilde{A}_y} = \frac{2P_0}{B_0 L}e^{2\tilde{A}_y/C_2}e^{2C_1\tilde{A}_x/(C_2C_3)} = \frac{2P_0}{B_0 L}\operatorname{sech}^2\tilde{z} = j_y \iff C_2 = \frac{2\mu_0 P_0}{B_0^2}.$$

4.4.2 Inverting the Weierstrass Transform

As aforementioned, we can solve the inverse problem exactly for the exponential functions in Eq. (4.14), using the fact that

$$\text{“}g_{js}(p_{js}) \propto \exp(\tilde{p}_{js})\text{''} \implies \text{“}P_j \propto \exp(\tilde{A}_j)\text{''},$$

with the terminology of Chap. 2. Hence the challenge is to try to solve

$$\tanh(\tilde{A}_x/C_3) = \frac{1}{\sqrt{2\pi}}\int_{-\infty}^{\infty} \exp\left[-\frac{1}{2}\left(\tilde{p}_{xs} - \frac{\operatorname{sgn}(q_s)}{\delta_s}\tilde{A}_x\right)^2\right] G_s(\tilde{p}_{xs})d\tilde{p}_{xs}. \qquad (4.15)$$

for some unknown G_s function, one component of a DF of the form

$$f_s = \frac{n_{0s}}{(\sqrt{2\pi}\,v_{th,s})^3}e^{-\beta_s H_s}\left(a_{0s}e^{\beta_s u_{xs} p_{xs}}e^{\beta_s u_{ys} p_{ys}} + a_{1s}G_s(p_{xs}) + b_s\right), \qquad (4.16)$$

and such that the species-dependent constants are yet to be determined. It turns out that Eq. (4.15) is not amenable to the Fourier transform method described in Sect. 1.3.5.5 since there does not exist an analytic expression for the Fourier transform of the tanh function. Furthermore, one cannot use the Hermite polynomial expansion techniques as developed in Chap. 2, because the Maclaurin expansion for $\tanh x$,

$$\tanh x = \sum_{n=0}^{\infty} \chi_n x^n,$$

is only convergent for $|x| < \pi/2$. This is not a purely formal objection, for the following reason. Using the theory developed in Chap. 2, we could in principle construct a Hermite polynomial expansion for the G_s function of the form

$$G_s = \sum_{n=0}^{\infty} \chi_n \mathrm{sgn}(q_s)^n \left(\frac{\delta_s}{\sqrt{2}}\right)^n H_n\left(\frac{p_{xs}}{\sqrt{2}m_s v_{\mathrm{th},s}}\right),$$

such that the the Weierstrass transform resulted in a Maclaurin series with the correct coefficients, χ_n. However, the Hermite series is valid for all p_{xs}—assuming that it is convergent—and there is a priori no reason to restrict the range of the conjugate variable, A_x. Hence the result of the forward procedure is a pressure function that is not convergent for all $\tilde{A}_x$, and cannot equal the closed form on the LHS of Eq. (4.15). Furthermore, since $\tilde{A}_x/C_3 = \tilde{z} \in (-\infty, \infty)$, one can not even make an argument on the basis of *accessibility* (i.e. claiming that this formal argument does not matter), which could possibly be justified if it were the case that $|\tilde{A}_x(\tilde{z})/C_3| < \pi/2 \,\forall \tilde{z}$. In the absence of other analytical techniques, one must proceed with this problem numerically. We do not develop that approach in detail in this thesis, but we shall show some indicative results, to demonstrate the principle.

In collaboration with J. D. B. Hodgson (who has led this particular effort), we have used *Genetic algorithms* (e.g. see Holland 1975) to construct numerical solutions for the G_s function. My contribution to this project has been on the theoretical side, whereas J.D.B. Hodgson's has been the development of the algortithm and numerical approach, as well as Figs. 4.4 and 4.5. The algorithm works by optimisation through random *mutation*. One starts with an initial *population* of candidate solutions to a problem, i.e. candidate G_s functions that could solve Eq. (4.15). Each member of the population (or *chromosome*) is ranked according to some *fitness* function. The population is then evolved in discrete steps (*generations*), between which various mutations and *genetic operations* occur, such that the fitness is hopefully optimised as $t \to \infty$.

Since the aim of the algorithm is—in general terms—to find a function $\mathcal{G}(p)$ that satisfies,

$$P(A) = \int_a^b K(A, p)\mathcal{G}(p)dp,$$

for known $P(A)$ and $K(A, p)$, a sensible fitness function to choose is

$$F(\mathcal{G}(p)) = \int_{A_0}^{A_1} \left[\int_a^b K(A, p)\mathcal{G}(p)dp \; - P(A)\right]^2 dA.$$

In analytic terms, one would of course use $\pm\infty$ for all the relevant integral limits, but clearly one cannot do this in numerical computation. Figure 4.4 displays some results for a run of the algorithm through 1000 generations. Figure 4.4a, b, and c display the highest ranked chromosome of each population at the initial, 4th, and 999th generations respectively. The highest ranked chromosome is the individual that best

Fig. 4.4 The ‘most fit’ numerical solution for the G_s function at three separate generations (courtesy of J.D.B. Hodgson)

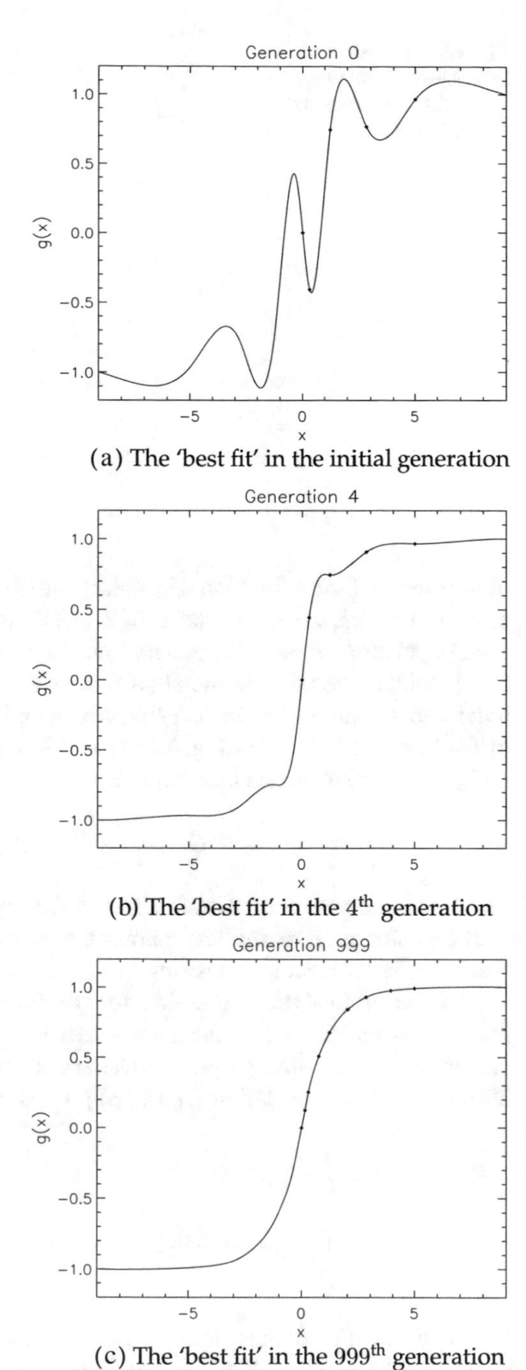

(a) The 'best fit' in the initial generation

(b) The 'best fit' in the 4th generation

(c) The 'best fit' in the 999th generation

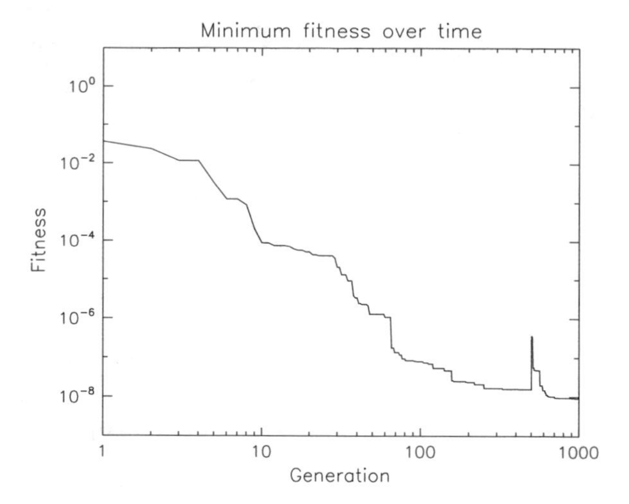

Fig. 4.5 The minimum fitness/error through the generations (courtesy of J.D.B. Hodgson)

minimises the fitness function (Eq. 4.4.2), which can be thought of as minimising the error. In Fig. 4.5, we show—on a loglog plot—the trend of the minimum fitness of each population, through the generations. The jump in the fitness around generation 500 identifies the point in the algorithm at which the grid resolution is increased, temporarily resulting in a larger error, which rapidly stabilises. An interesting feature of the 'solution' given by Fig. 4.4c is that it almost directly lies over the function $\tanh(\tilde{p}_{xs}/C_3)$, and hence it seems that

$$G_s(\tilde{p}_{xs}) \approx \tanh(\tilde{p}_{xs}/C_3).$$

The numerical procedure therefore seems to suggest that $\tanh x$ is close to a 'numerical' eigenfunction of the Weierstrass transform, despite the fact that one cannot compute the Weierstrass transform of the tanh function.

Without an analytic expression for the function $G_s(\tilde{p}_{xs})$, we can make some progress in understanding the micro-macroscopic parameter relationships, and in calculating the bulk flow properties. Using standard integrals (Gradshteyn and Ryzhik 2007), we see that the DF in Eq. (4.16) gives a pressure tensor of the form

$$\begin{aligned} P_{zz} &= \sum_s m_s \int v_z^2 f_s d^3 v, \\ &= \sum_s \frac{n_{0s}}{\beta_s}\left(a_{0s} e^{(u_{xs}^2+u_{ys}^2)/(2v_{th,s}^2)} e^{\beta_s u_{xs} q_s A_x} e^{\beta_s u_{ys} q_s A_y} + a_{1s}\tanh(\tilde{A}_x/C_3) + b_s\right). \end{aligned}$$

Channell's method dictates that this expression must match up with the macroscopic expression from Eq. (4.14). This condition, as well as that of imposing $\sigma = 0$ gives the following conditions

$$\frac{2C_1}{C_2C_3B_0L} = e\beta_i u_{xi} = -e\beta_e u_{xe}, \tag{4.17}$$
$$\frac{2}{C_2B_0L} = e\beta_i u_{yi} = -e\beta_e u_{ye},$$

$$n_{0s}a_{0s}e^{(u_{xs}^2+u_{ys}^2)/(2v_{th,s}^2)} =: a_0 = \frac{\beta_e\beta_i}{\beta_e+\beta_i}P_0C_2, \tag{4.18}$$

$$n_{0s}a_{1s} =: a_1 = -2\frac{\beta_e\beta_i}{\beta_e+\beta_i}P_0C_1, \tag{4.19}$$

$$n_{0s}b_s =: b = \frac{\beta_e\beta_i}{\beta_e+\beta_i}P_0C_b,$$

and, for completeness, the number density is given by

$$n_i = n_e := n = a_0\text{sech}^2\tilde{z} + a_1\tanh\tilde{z} + b = \frac{\beta_e\beta_i}{\beta_e+\beta_i}P_{zz}.$$

The conditions listed above represent 10 constraints for 14 parameters (β_s, n_{0s}, u_{xs}, u_{ys}, a_{0s}, a_{1s}, b_s), given macroscopic characteristics B_0, P_0, C_1, C_2, C_3, C_b and L.

We can also calculate the bulk flow properties. In particular one should check that $j_x = 0$. Using standard integrals (Gradshteyn and Ryzhik 2007), we see that

$$\begin{aligned} j_x = 0 &= \sum_s q_s \int f_s v_x d^3v, \\ &= \sum_s q_s n_{0s}\Bigg[a_{0s}u_{xs}e^{(u_{xs}^2+u_{ys}^2)/(2v_{th,s}^2)}e^{\beta_s q_s u_{ys}A_y}e^{\beta_s q_s u_{xs}A_x} \\ &+ \frac{a_{1s}}{\sqrt{2\pi}\,v_{th,s}}\int_{-\infty}^{\infty} e^{-v_x^2/(2v_{th,s}^2)}v_x G_s(\tilde{p}_{xs})dv_x\Bigg]. \end{aligned}$$

By differentiating Eq. (4.15) with respect to $\tilde{A}_x$, we can see that

$$\int_{-\infty}^{\infty} e^{-v_x^2/(2v_{th,s}^2)}v_x G_s(\tilde{p}_{xs})dv_x = \frac{\sqrt{2\pi}\,\delta_s\text{sgn}(q_s)v_{th,s}^2}{C_3}\text{sech}^2(\tilde{A}_x/C_3). \tag{4.20}$$

Plugging this back into the equation for j_x gives

$$j_x = \sum_s q_s\left[a_0u_{xs}\text{sech}^2\tilde{z} + \frac{a_1}{C_3\beta_s q_s B_0L}\text{sech}^2\tilde{z}\right], \tag{4.21}$$

and substituting in Eq. (4.17), and then Eqs. (4.18) and (4.19) gives

$$j_x = \frac{\beta_e\beta_i\text{sech}^2\tilde{z}}{(\beta_e+\beta_i)C_3B_0L}\sum_s\frac{1}{\beta_s}\left[2C_1P_0 - 2P_0C_1\right] = \frac{\text{sech}^2\tilde{z}P_0}{C_3B_0L}\sum_s 0 = 0. \tag{4.22}$$

In contrast to the solution found by Alpers (1969), we see that this DF gives $V_{xs} \propto j_x = 0$.

Similarly, we can calculate j_y,

$$j_y = \sum_s q_s \int f_s v_y d^3 v,$$

$$\vdots$$

$$= \frac{C_2 B_0^2}{2\mu_0} \frac{2}{B_0 L} \text{sech}^2 \tilde{z} = j_y.$$

The individual bulk velocities in the y direction are proportional to the total current density, and go to zero at ∞, i.e. $V_{ys} \propto j_y$.

4.5 The Analytical/"Exponential" Equilibrium DF

4.5.1 The Pressure Tensor

In this section we derive one more pressure tensor consistent with the AH+G field, that allows an exact analytical solution for the DF. The key step for analytic progress is to find distinct representations of $\tanh \tilde{z} = \tanh \tilde{z}(\tilde{A}_x, \tilde{A}_y)$ that allow inversion of the Weierstrass transform.

In a similar vein to the method in Alpers (1969), we achieve this crucial step by identifying two distinct representations of $\tanh \tilde{z}(A_x, A_y)$,

$$\tanh \tilde{z} = 1 - e^{-\tilde{z}} \text{sech} \tilde{z} = 1 - e^{\frac{C_1 - C_2}{C_2 C_3} \tilde{A}_x} e^{\frac{1}{C_2} \tilde{A}_y},$$

$$\tanh \tilde{z} = \sqrt{1 - \text{sech}^2 \tilde{z}} = \sqrt{1 - e^{\frac{2C_1}{C_2 C_3} \tilde{A}_x} e^{\frac{2}{C_2} \tilde{A}_y}},$$

These are composed as a linear combination, and then substituted into Eq. (4.9) to give

$$P_{zz} = P_T - \frac{B_0^2}{2\mu_0} \Bigg\{ C_1^2 + C_3^2 + 2C_1 C_2 \left(1 - e^{\frac{C_1 - C_2}{C_2 C_3} \tilde{A}_x} e^{\frac{1}{C_2} \tilde{A}_y}\right)$$
$$+ C_2^2 \left[k \left(1 - e^{\frac{C_1 - C_2}{C_2 C_3} \tilde{A}_x} e^{\frac{1}{C_2} \tilde{A}_y}\right)^2 + (1-k) \left(1 - e^{\frac{2C_1}{C_2 C_3} \tilde{A}_x} e^{\frac{2}{C_2} \tilde{A}_y}\right) \right] \Bigg\}, \quad (4.23)$$

with k a 'separation constant'. Ampère's Law implies that P_{zz} must satisfy $\partial P_{zz} / \partial A_x(\tilde{z}) = 0$ and $\partial P_{zz} / \partial A_y(\tilde{z}) = B_0 C_2 / (\mu_0 L) \text{sech}^2 \tilde{z}$, and it is seen to do so when $k = C_1 / C_2$. In this case, Eq. (4.23) can be re-written

$$P_{zz} = P_T - \frac{B_0^2}{2\mu_0}\left\{C_1^2 + C_3^2 + -C_1C_2 + C_1C_2\left(1 + \left(1 - e^{\frac{C_1-C_2}{C_2C_3}\tilde{A}_x} e^{\frac{1}{C_2}\tilde{A}_y}\right)\right)^2\right.$$

$$\left. + C_2(C_2 - C_1)\left(1 - e^{\frac{2C_1}{C_2C_3}\tilde{A}_x} e^{\frac{2}{C_2}\tilde{A}_y}\right)\right\}, \tag{4.24}$$

An examination of the coefficients of the exponential functions in Eq. (4.24) tells us that $P_{zz} > 0 \,\forall\, (A_x, A_y)$ under the following conditions

$$C_1C_2 < 0, \tag{4.25}$$

$$P_T > \frac{B_0^2}{2\mu_0}\left[C_1^2 + C_3^2 - C_1C_2 + C_2^2 - C_1C_2\right]$$

$$= \frac{B_0^2}{2\mu_0}\left[C_1^2 + C_2^2 + C_3^2 - 2C_1C_2\right] \tag{4.26}$$

Now that a $P_{zz} > 0$ has been found that satisfies Ampère's Law and pressure balance, we can attempt to solve the inverse problem.

4.5.2 The DF

By comparison with Eq. (4.24) (in which P_{zz} is written as a sum of exponential functions), we can suggest a form for the DF by using either 'inspection and standard integral formulae' (Gradshteyn and Ryzhik 2007), Fourier transforms (see Sect. 1.3.5.5), or knowledge of eigenfunctions (Wolf 1977). The form that we choose is

$$f_s = \frac{n_{0s}}{(\sqrt{2\pi}\, v_{\mathrm{th},s})^3} e^{-\beta_s H_s}\left(a_{0s} e^{\beta_s(u_{xs}p_{xs} + u_{ys}p_{ys})}\right.$$

$$\left. + a_{1s} e^{2\beta_s(u_{xs}p_{xs} + u_{ys}p_{ys})} + a_{2s} e^{\beta_s(v_{xs}p_{xs} + v_{ys}p_{ys})} + b_s\right), \tag{4.27}$$

for $a_{0s}, a_{1s}, a_{2s}, b_s, u_{xs}, u_{ys}, v_{xs}$ and v_{ys} as yet arbitrary constants, with the "a, b" constants dimensionless, and the "u, v" constants the bulk flows of particular particle populations (e.g. see Davidson 2001; Schindler 2007 and Sect. 1.3.2.2).

4.5.2.1 Equilibrium Parameters and Their Relationships

We proceed with the necessary task of ensuring that the DF in Eq. (4.27) exactly reproduces the correct pressure tensor expression of Eq. (4.23). After some algebra we find the 'micro-macroscopic' consistency relations by taking the v_z^2 moment of the DF, and these are displayed in Eqs. (4.28–4.31).

$$P_T - \frac{B_0^2}{2\mu_0}\left[(C_1+C_2)^2 + C_3^2\right] = b\frac{\beta_e+\beta_i}{\beta_e\beta_i}, \quad \frac{C_1-C_2}{C_2C_3B_0L} = e\beta_i u_{xi} = -e\beta_e u_{xe}, \tag{4.28}$$

$$4C_1C_2\frac{B_0^2}{2\mu_0} = a_0\frac{\beta_e+\beta_i}{\beta_e\beta_i}, \quad \frac{1}{C_2B_0L} = e\beta_i u_{yi} = -e\beta_e u_{ye}, \tag{4.29}$$

$$-C_1C_2\frac{B_0^2}{2\mu_0} = a_1\frac{\beta_e+\beta_i}{\beta_e\beta_i}, \quad \frac{2C_1}{C_2C_3B_0L} = e\beta_i v_{xi} = -e\beta_e v_{xe}, \tag{4.30}$$

$$C_2(C_2-C_1)\frac{B_0^2}{2\mu_0} = a_2\frac{\beta_e+\beta_i}{\beta_e\beta_i}, \quad \frac{2}{C_2B_0L} = e\beta_i v_{yi} = -e\beta_e v_{ye}, \tag{4.31}$$

We must also ensure that $n_i(A_x, A_y) = n_e(A_x, A_y)$ (for $n_s(A_x, A_y)$ the number density of species s) in order to be consistent with our assumption that $\phi = 0$. The constants a_0, a_1, a_2 and b are defined by these neutrality relations that complete this final step of the method, are found by calculating the zeroth order moment of the DF, and are written in Eqs. (4.32 and 4.33).

$$a_0 = n_{0s}a_{0s}e^{(u_{xs}^2+u_{ys}^2)/(2v_{\text{th},s}^2)}, \quad a_2 = n_{0s}a_{2s}e^{(v_{xs}^2+v_{ys}^2)/(2v_{\text{th},s}^2)}, \tag{4.32}$$

$$a_1 = n_{0s}a_{1s}e^{2(u_{xs}^2+u_{ys}^2)/v_{\text{th},s}^2}, \quad b = n_{0s}b_s. \tag{4.33}$$

These constraints are 16 in number, with 20 microscopic parameters (β_s, n_{0s}, a_{0s}, a_{1s}, a_{2s}, b_s, u_{xs}, u_{ys}, v_{xs}, v_{ys}), given chosen macroscopic parameters (B_0, P_T, L, C_1, C_2, C_3).

4.5.2.2 Non-negativity of the DF

Since we integrate f_s over velocity space to calculate P_{zz}, it is clear that non-negativity of P_{zz} does not imply non-negativity of f_s. Furthermore, it is clear from Eqs. (4.29) and (4.32) that $C_1C_2 < 0 \implies a_{0s} < 0$ (as well as $a_{1s} > 0$, $a_{2s} > 0$). We can also see by consideration of Eqs. (4.26) and (4.33) that b_s>0. The fact that $a_{0s} < 0$ is a cause for concern, regarding the positivity of the DF, given its form (Eq. 4.27). However, by completing the square, the DF can be re-written as

$$f_s = \frac{n_{0s}}{(\sqrt{2\pi}\, v_{\text{th},s})^3}e^{-\beta_s H_s}\left[\frac{1}{a_{1s}}\left(-\frac{a_{0s}}{2} + a_{1s}e^{\beta_s(u_{xs}p_{xs}+u_{ys}p_{ys})}\right)^2 - \frac{a_{0s}^2}{4a_{1s}}\right.$$
$$\left. + a_{2s}e^{\beta_s(v_{xs}p_{xs}+v_{ys}p_{ys})} + b_s\right].$$

Hence we see that non-negativity of the DF is assured provided

$$b_s \geq \frac{a_{0s}^2}{4a_{1s}}. \tag{4.34}$$

4.5.2.3 The DF Is a Sum of Maxwellians

The equilibrium DF in Eq. (4.27) is written as a function of the constants of motion (H_s, p_{xs}, p_{ys}), and this was suitable for constructing an exact equilibrium solution to the Vlasov equation. However, we can write f_s explicitly as a function over phase-space $(z, \boldsymbol{v})$, in a form similar to that of the drifting Maxwellian in Eq. (3.47). The DF can be re-written as

$$f_s(z, \boldsymbol{v}) = \frac{1}{(\sqrt{2\pi}\, v_{\text{th},s})^3}\Bigg[\mathcal{N}_{0s}(z)e^{-\frac{(\boldsymbol{v}-\boldsymbol{V}_{0s})^2}{2v_{\text{th},s}^2}} + \mathcal{N}_{1s}(z)e^{-\frac{(\boldsymbol{v}-\boldsymbol{V}_{1s})^2}{2v_{\text{th},s}^2}} \\ + \mathcal{N}_{2s}(z)e^{-\frac{(\boldsymbol{v}-\boldsymbol{V}_{2s})^2}{2v_{\text{th},s}^2}} + be^{-\frac{v^2}{2v_{\text{th},s}^2}}\Bigg], \tag{4.35}$$

for the density and bulk flow variables ("$\mathcal{N}$, $\boldsymbol{V}$"), defined by

$$\begin{aligned}
\mathcal{N}_{0s}(z) &= a_0 e^{q_s\beta_s \boldsymbol{A}\cdot\boldsymbol{V}_{0s}} = a_0 e^{-\tilde{z}}\text{sech}\tilde{z}, \quad \boldsymbol{V}_{0s} = (u_{xs}, u_{ys}, 0),\\
\mathcal{N}_{1s}(z) &= a_1 e^{q_s\beta_s \boldsymbol{A}\cdot\boldsymbol{V}_{1s}} = a_1 e^{-2\tilde{z}}\text{sech}^2\tilde{z}, \quad \boldsymbol{V}_{1s} = (2u_{xs}, 2u_{ys}, 0),\\
\mathcal{N}_{2s}(z) &= a_2 e^{q_s\beta_s \boldsymbol{A}\cdot\boldsymbol{V}_{2s}} = a_2\text{sech}^2\tilde{z}, \quad \boldsymbol{V}_{2s} = (v_{xs}, v_{ys}, 0),
\end{aligned}$$

respectively. The u, v variables are normalised by $v_{\text{th},s}$ ($\tilde{u}_{xs} = u_{xs}/v_{\text{th},s}$ etc.). This representation of f_s has the advantages of having a clear visual/physical interpretation, and of being in a form readily implemented into PIC simulations as initial conditions. Despite the fact that each term of f_s as written in Eq. (4.35) bears a strong resemblance to $f_{Maxw,s}$ as defined by Eq. (3.47), f_s is an exact Vlasov equilibrium DF, whereas $f_{Maxw,s}$ is not.

4.5.3 *Plots of the DF*

In order to plot the normalised DF, $\tilde{f}_s = f_s/\max f_s$, it is more convenient for Eqs. (4.28)–(4.31) to be expressed in dimensionless form. Making use of the dimensionless parameters also defined in Sect. 4.5.2.3, we have the following relationships

$$\begin{aligned}
\beta_T - \left[(C_1+C_2)^2 + C_3^2\right] &= bR, & \frac{(C_1-C_2)\delta_s^\star}{C_2C_3} &= \tilde{u}_{xs}\\
4C_1C_2 &= a_0R, & \frac{\delta_s^\star}{C_2} &= \tilde{u}_{ys},\\
-C_1C_2 &= a_1R, & \frac{2C_1\delta_s^\star}{C_2C_3} &= \tilde{v}_{xs},\\
C_2(C_2-C_1) &= a_2R, & \frac{2\delta_s^\star}{C_2} &= \tilde{v}_{ys}.
\end{aligned}$$

The signed magnetisation parameter $\delta_s^\star = m_s v_{\mathrm{th},s}/(q_s B_0 L)$ is the ratio of the (signed) thermal Larmor radius to the current sheet width, and the constants R and β_T defined by

$$R = \frac{\beta_e + \beta_i}{\beta_e \beta_i} \frac{2\mu_0}{B_0^2},$$
$$\beta_T = P_T \frac{2\mu_0}{B_0^2}.$$

Hence, the normalised bulk flow parameters, $\tilde{u}_{xs}, \tilde{u}_{ys}, \tilde{v}_{xs}, \tilde{v}_{ys}$ are fixed by choosing the magnetisation, $\delta_s^\star$, and the magnetic field configuration, C_1, C_2, C_3. If in addition one chooses n_{0s}, and the ratio R (note that n_{0s} R is dimensionless), then we see that the 'density parameters' a_0, a_1 and a_2 are also fixed. In turn a_{0s}, a_{1s} and a_{2s} are then fixed by Eqs. (4.32) and (4.33). Then, the lower bound on b_s (for positivity of the DF) is determined by Eq. (4.34), and in turn we see a lower bound for b and hence β_T.

Note that when $T_e = T_i := T$ and $n_{0i} = n_{0e} := n_0$, it is the case that $n_0 R = 2\beta_{pl}^\star$, for

$$\beta_{pl}^\star = \frac{n_0 k_B T}{B_0^2/(2\mu_0)},$$

a constant reference value for β_{pl}, which itself is spatially dependent. We shall also assume that $b_s = a_{0s}^2/(4a_{1s})$, and hence

$$\inf \tilde{f}_s = 0.$$

In Fig. 4.6 we present plots of the DF in $(v_x/v_{\mathrm{th},s}, v_y/v_{\mathrm{th},s})$ space, for $z/L = (0, 0.1, 1, 10)$, and for the parameters

$$(\delta_i, R, n_0, C_1, C_2, C_3) = (0.2, 0.1, 1, -0.1, 0.2, 0.1),$$
$$\Longrightarrow (\tilde{u}_{xi}, \tilde{u}_{yi}, \tilde{v}_{xi}, \tilde{v}_{yi}) = (-3, 1, -2, 2).$$

We have chosen this particular parameter set, in order to clearly see that the VM equilibrium permits multiple maxima in velocity space, as is to be expected by a sum of drifting Maxwellians. However, whilst the plots of $\tilde{f}_i$ permit multiple maxima for $z/L = 0, 0.1, 1$ in the parameter range chosen, we see that for large z/L the DF is an isotropic Maxwellian, centred on (0, 0). This is consistent with no bulk flows V_{xi}, V_{xe} for large $\tilde{z}$, in contrast to the DF found by Alpers (1969).

In particular, Fig. 4.6e shows $\tilde{f}_e$ for $\delta_e = \delta_i$, and hence

$$(\tilde{u}_{xe}, \tilde{u}_{ye}, \tilde{v}_{xe}, \tilde{v}_{ye}) = -(\tilde{u}_{xi}, \tilde{u}_{yi}, \tilde{v}_{xi}, \tilde{v}_{yi}),$$

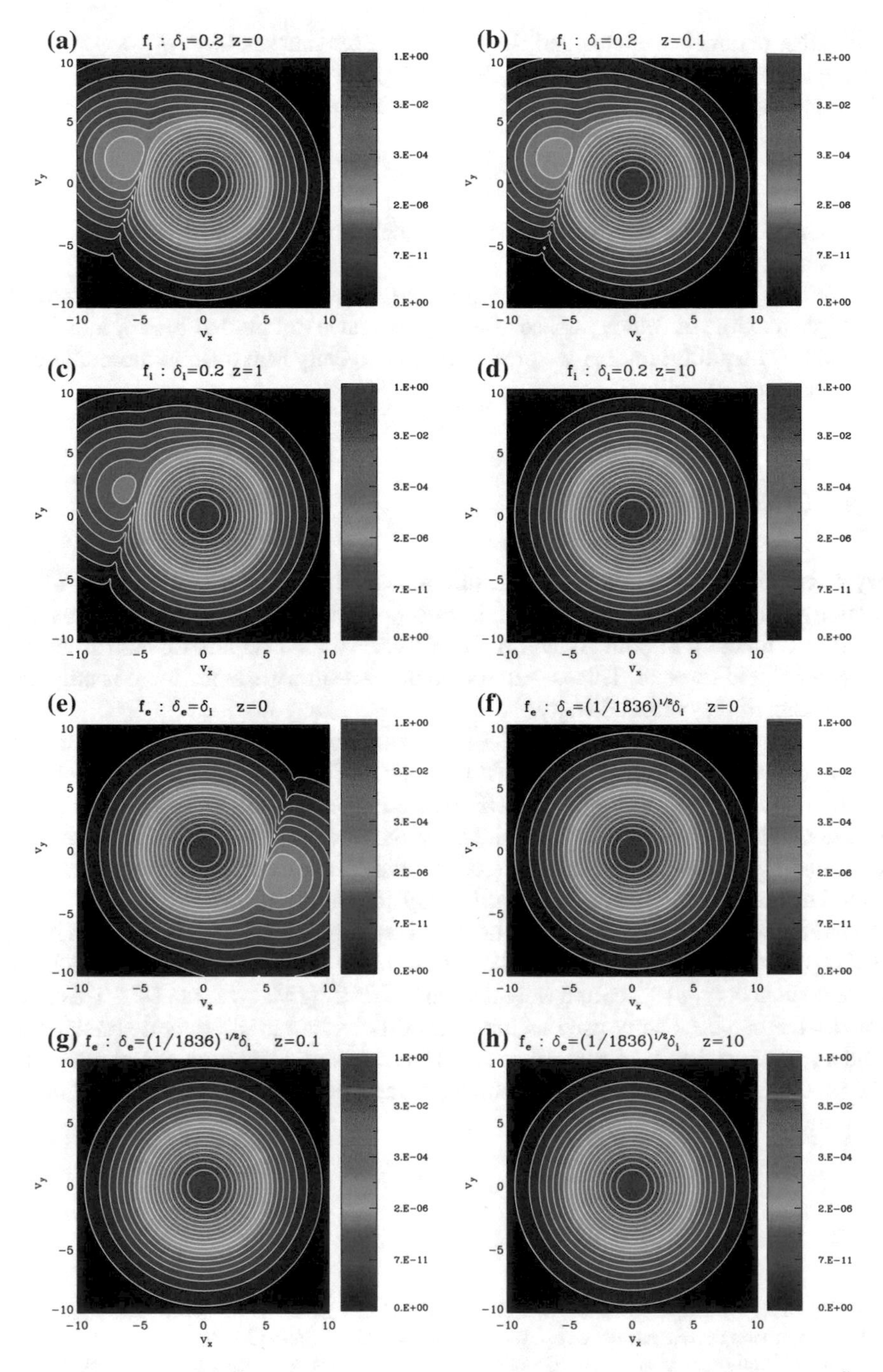

Fig. 4.6 In **a–d** we plot $\tilde{f}_i$ for $\delta_i = 0.2$ and $z/L = 0, 0.1, 1$ and 10 respectively. In **e** we plot $\tilde{f}_e$ for $\delta_e = 0.2$ and $z/L = 0$. In **f–h** we plot $\tilde{f}_e$ for $\delta_e = \sqrt{m_e/m_i}\delta_i$ and $z/L = 0, 0.1$ and 10 respectively

with other parameters unchanged. As a result, we see that sending "$q_i \rightarrow q_e$" seems equivalent to sending "$f_i(v_x/v_{\text{th},i}, v_y/v_{\text{th},i}) \rightarrow f_e(-v_x/v_{\text{th},e}, -v_y/v_{\text{th},e})$". However, for Fig. 4.6f, g and h we take $T_e = T_i$, and hence $\delta_e = \sqrt{m_e/m_i}\delta_i$, giving

$$(\tilde{u}_{xe}, \tilde{u}_{ye}, \tilde{v}_{xe}, \tilde{v}_{ye}) \approx (0.07, -0.02, 0.05, -0.05).$$

The normalised bulk electron flow is now much smaller in magnitude, and this is represented in the figures.

We note that there is a large portion of parameter space for which one sees no multiple maxima in velocity space (although we have not plotted these), indicating that the VM equilibrium that we present permits locally Maxwellian/thermalised—and hence micro-stable -DFs.

4.6 Discussion

By considering the theory of the pressure tensor in vector-potential space (and its analogy with the problem of a particle in a potential), we have deduced that P_{zz} must be a function of both A_x and A_y, to describe a 1D asymmetric Harris current sheet with field reversal. This is—at first glance—a surprise, since there is only one component of the current density.

We have presented two valid $P_{zz}(A_x, A_y)$ functions that are self-consistent with an asymmetric Harris sheet plus guide field. One of these necessitated a numerical approach in order to solve for the DF, whereas the second allowed an analytical solution. The magnetic fields described by our models have often been used as asymmetric current sheet models for reconnection studies, and should be particularly suited to studying reconnection in Earth's dayside magnetopause.

The expression for the exact analytical VM equilibrium DF is elementary in form, and is written as a sum of exponential functions of the constants of motion, which can be re-written in $(z, \boldsymbol{v})$ space as a weighted sum of drifting Maxwellian DFs. This form for the DF can be readily used as initial conditions in particle-in-cell simulations. The equilibrium has zero mass flow far from the sheet, which is corroborated by the plots of the DF, and this is in contrast to the known exact analytical DF in the literature (Alpers 1969).

References

O. Allanson, F. Wilson, T. Neukirch, Y.-H. Liu, J.D.B. Hodgson, Exact Vlasov-Maxwell equilibria for asymmetric current sheets. Geophys. Res. Lett. **44**, 8685–8695 (2017)

W. Alpers, Steady state charge neutral models of the magnetopause. Astrophys. Space Sci. **5**, 425–437 (1969)

N. Aunai, G. Belmont, R. Smets, First demonstration of an asymmetric kinetic equilibrium for a thin current sheet. Phys. Plasmas **20**(11), 110702 (2013a)

N. Aunai, M. Hesse, S. Zenitani, M. Kuznetsova, C. Black, R. Evans, R. Smets, Comparison between hybrid and fully kinetic models of asymmetric magnetic reconnection: coplanar and guide field configurations. Phys. Plasmas **20**(2), 022902 (2013b)

G. Belmont, N. Aunai, R. Smets, Kinetic equilibrium for an asymmetric tangential layer. Phys. Plasmas **19**(2), 022108 (2012)

J.L. Burch, R.B. Torbert, T.D. Phan, L.-J. Chen, T.E. Moore, R.E. Ergun, J.P. Eastwood, D.J. Gershman, P.A. Cassak, M.R. Argall, S. Wang, M. Hesse, C.J. Pollock, B.L. Giles, R. Nakamura, B.H. Mauk, S.A. Fuselier, C.T. Russell, R.J. Strangeway, J.F. Drake, M.A. Shay, Y.V. Khotyaintsev, P.-A. Lindqvist, G. Marklund, F.D. Wilder, D.T. Young, K. Torkar, J. Goldstein, J.C. Dorelli, L.A. Avanov, M. Oka, D.N. Baker, A.N. Jaynes, K.A. Goodrich, I.J. Cohen, D.L. Turner, J.F. Fennell, J.B. Blake, J. Clemmons, M. Goldman, D. Newman, S.M. Petrinec, K.J. Trattner, B. Lavraud, P.H. Reiff, W. Baumjohann, W. Magnes, M. Steller, W. Lewis, Y. Saito, V. Coffey, M. Chandler, Electron-scale measurements of magnetic reconnection in space. Science **352**, aaf2939 (2016)

J.L. Burch, T.D. Phan, Magnetic reconnection at the dayside magnetopause: advances with MMS. Geophys. Res. Lett. **43**, 8327–8338 (2016)

P.A. Cassak, M.A. Shay, Scaling of asymmetric magnetic reconnection: general theory and collisional simulations. Phys. Plasmas **14**(10), 102114 (2007)

P.J. Channell, Exact Vlasov-Maxwell equilibria with sheared magnetic fields. Phys. Fluids **19**, 1541–1545 (1976)

J. Dargent, N. Aunai, G. Belmont, N. Dorville, B. Lavraud, M. Hesse, Full particle-in-cell simulations of kinetic equilibria and the role of the initial current sheet on steady asymmetric magnetic reconnection. J. Plasma Phys. **82**(3), 905820305 (2016)

R.C. Davidson, *Physics of Nonneutral Plasmas* (World Scientific Press, 2001)

N. Dorville, G. Belmont, N. Aunai, J. Dargent, L. Rezeau, Asymmetric kinetic equilibria: generalization of the BAS model for rotating magnetic profile and non-zero electric field. Phys. Plasmas **22**(9), 092904 (2015)

J.T. Gosling, S. Eriksson, R.M. Skoug, D.J. McComas, R.J. Forsyth, Petschek-type reconnection exhausts in the solar wind well beyond 1 AU: ulysses. Astrophys. J. **644**, 613–621 (2006)

I.S. Gradshteyn, I.M. Ryzhik, *Table of Integrals, Series, and Products*, 7th edn. (Elsevier/Academic Press, Amsterdam, 2007), 1171 pages

M.A. Hapgood, Space physics coordinate transformations-a user guide. Planet. Space Sci. **40**, 711–717 (1992)

E.G. Harris, On a plasma sheath separating regions of oppositely directed magnetic field. Nuovo Cimento **23**, 115 (1962)

M. Hesse, N. Aunai, S. Zenitani, M. Kuznetsova, J. Birn, Aspects of collisionless magnetic reconnection in asymmetric systems. Phys. Plasmas **20**(6), 061210 (2013)

M. Hesse, N. Aunai, D. Sibeck, J. Birn, On the electron diffusion region in planar, asymmetric, systems. Geophys. Res. Lett. **41**, 8673–8680 (2014)

J.H. Holland, *Adaptation in Natural and Artificial Systems: An Introductory Analysis with Applications to Biology, Control, and Artificial Intelligence* (University of Michigan Press, 1975)

J. Huang, Z.W. Ma, D. Li, Debye-length scaled structure of perpendicular electric field in collisionless magnetic reconnection. Geophys. Res. Lett. **35**(10) (2008)

B.B. Kadomtsev, Disruptive instability in Tokamaks. Sov. J. Plasma Phys. **1**, 710–715 (1975)

J.R. Kan, Equilibrium configurations of Vlasov plasmas carrying a current component along an external magnetic field. J. Plasma Phys. **7**, 445–459 (1972)

H. Karimabadi, V. Roytershteyn, W. Daughton, Y.-H. Liu, Recent evolution in the theory of magnetic reconnection and its connection with turbulence. Space Sci. Rev. **178**, 307–323 (2013)

S. Kobayashi, B.N. Rogers, R. Numata, Gyrokinetic simulations of collisionless reconnection in turbulent non-uniform plasmas. Phys. Plasmas **21**(4), 040704 (2014)

C.M. Komar, R.L. Fermo, P.A. Cassak, Comparative analysis of dayside magnetic reconnection models in global magnetosphere simulations. J. Geophys. Res. (Space Phys.) **120**, 276–294 (2015)

M.M. Kuznetsova, M. Roth, Thresholds for magnetic percolation through the magnetopause current layer in asymmetrical magnetic fiels. J. Geophys. Res. **100**, 155–174 (1995)
L.C. Lee, J.R. Kan, A unified kinetic model of the tangential magnetopause structure. J. Geophys. Res. (Space Phys.) **84**, 6417–6426 (1979)
J. Lemaire, L.F. Burlaga, Diamagnetic boundary layers-a kinetic theory. Astrophys. Space Sci. **45**, 303–325 (1976)
M.G. Linton, Reconnection of nonidentical flux tubes. J. Geophys. Res. (Space Phys.) **111**, A12S09 (2006)
Y.-H. Liu, M. Hesse, Suppression of collisionless magnetic reconnection in asymmetric current sheets. Phys. Plasmas **23**(6), 060704 (2016)
Y.-H. Liu, M. Hesse, M. Kuznetsova, Orientation of X lines in asymmetric magnetic reconnection-mass ratio dependency. J. Geophys. Res. (Space Phys.) **120**, 7331–7341 (2015)
K. Malakit, M.A. Shay, P.A. Cassak, C. Bard, Scaling of asymmetric magnetic reconnection: kinetic particle-in-cell simulations. J. Geophys. Res. (Space Phys.) **115**, A10223 (2010)
N.A. Murphy, M.P. Miralles, C.L. Pope, J.C. Raymond, H.D. Winter, K.K. Reeves, D.B. Seaton, A.A. van Ballegooijen, J. Lin, Asymmetric magnetic reconnection in solar flare and coronal mass ejection current sheets. Astrophys. J. **751**(56), 56 (2012)
M. Øieroset, T.D. Phan, M. Fujimoto, Wind observations of asymmetric magnetic reconnection in the distant magnetotail. Geophys. Res. Lett. **31**, L12801 (2004)
T.D. Phan, M.A. Shay, J.T. Gosling, M. Fujimoto, J.F. Drake, G. Paschmann, M. Oieroset, J.P. Eastwood, V. Angelopoulos, Electron bulk heating in magnetic reconnection at earth's magnetopause: dependence on the inflow Alfvén speed and magnetic shear. Geophys. Res. Lett. **40**, 4475–4480 (2013)
P.L. Pritchett, Collisionless magnetic reconnection in an asymmetric current sheet. J. Geophys. Res. (Space Phys.) **113**, A06210 (2008)
M.J. Pueschel, P.W. Terry, D. Told, F. Jenko, Enhanced magnetic reconnection in the presence of pressure gradients. Phys. Plasmas **22**(6), 062105 (2015)
K.B. Quest, F.V. Coroniti, Tearing at the dayside magnetopause. J. Geophys. Res. **86**, 3289–3298 (1981)
M. Roth, J. de Keyser, M.M. Kuznetsova, Vlasov theory of the equilibrium structure of tangential discontinuities in space plasmas. Space Sci. Rev. **76**, 251–317 (1996)
V. Roytershteyn, W. Daughton, H. Karimabadi, F.S. Mozer, Influence of the lower-hybrid drift instability on magnetic reconnection in asymmetric configurations. Phys. Rev. Lett. **108**(18), 185001 (2012)
K. Schindler, *Physics of Space Plasma Activity* (Cambridge University Press, 2007)
S. Servidio, W.H. Matthaeus, M.A. Shay, P.A. Cassak, P. Dmitruk, Magnetic reconnection in two-dimensional magnetohydrodynamic turbulence. Phys. Rev. Lett. **102**(11), 115003 (2009)
M. Swisdak, B.N. Rogers, J.F. Drake, M.A. Shay. Diamagnetic suppression of component magnetic reconnection at the magnetopause. J. Geophys. Res. (Space Phys.) **108**, 1218 (2003)
L. Trenchi, M.F. Marcucci, R.C. Fear, The effect of diamagnetic drift on motion of the dayside magnetopause reconnection line. Geophys. Res. Lett. **42**, 6129–6136 (2015)
P.-R. Wang, C. Huang, Q.-M. Lu, R.-S. Wang, S. Wang, Numerical simulations of magnetic reconnection in an asymmetric current sheet. Chin. Phys. Lett. **30**(12), 125202 (2013)
K.B. Wolf, On self-reciprocal functions under a class of integral transforms. J. Math. Phys. **18**(5), 1046–1051 (1977)
L. Zakharov, B. Rogers, Two-fluid magnetohydrodynamic description of the internal kink mode in tokamaks. Phys. Fluids B **4**, 3285–3301 (1992)
C. Zhu, R. Liu, D. Alexander, X. Sun, R.T.J. McAteer, Complex flare dynamics initiated by a filament-filament interaction. Astrophys. J. **813**(60), 60 (2015)

Chapter 5
Neutral and Non-neutral Flux Tube Equilibria

Things are the way they are because they were the way they were.

Fred Hoyle

Much of the work in this chapter is drawn from Allanson et al. (2016).

5.1 Preamble

In this chapter we calculate exact 1D collisionless plasma equilibria for a continuum of flux tube models, for which the total magnetic field is made up of the 'force-free' Gold-Hoyle (GH) magnetic flux tube embedded in a uniform and anti-parallel background magnetic field. For a sufficiently weak background magnetic field, the axial component of the total magnetic field reverses at some finite radius. The presence of the background magnetic field means that the total system is not exactly force-free, but by reducing its magnitude, the departure from force-free can be made as small as desired. The DF for each species is a function of the three constants of motion; namely, the Hamiltonian and the canonical momenta in the axial and azimuthal directions. Poisson's equation and Ampère's law are solved exactly, and the solution allows either electrically neutral or non-neutral configurations, depending on the values of the bulk ion and electron flows. These equilibria have possible applications in various solar, space, and astrophysical contexts, as well as in the laboratory.

The work in this chapter pertains to a cylindrical geometry, in which r is the horizontal distance from the z axis, and θ the azimuthal angle.

O. Allanson, *Theory of One-Dimensional Vlasov-Maxwell Equilibria*, Springer Theses, https://doi.org/10.1007/978-3-319-97541-2_5

5.2 Introduction

Magnetic flux tubes and flux ropes are prevalent in the study of plasmas, with a wide variety of observed forms in nature and experiment, as well as uses and applications in numerical experiments and theory. Some examples of the environments and fields of study in which they feature include solar (e.g. Priest et al. 2002; Magara and Longcope 2003); solar wind (e.g. Wang and Sheeley 1990; Borovsky 2008); planetary magnetospheres (e.g. Sato et al. 1986; Pontius and Wolf 1990) and magnetopauses (e.g. Cowley and Owen 1989); astrophysical plasmas (e.g. Rogava et al. 2000; Li et al. 2006); tokamak (e.g. Bottino et al. 2007; Ham et al. 2016), laboratory pinch experiments (e.g. Rudakov et al. 2000), and the basic study of energy release in magnetised plasmas (e.g. Cowley et al. 2015), to give a small selection of references.

One application of flux tubes is in the study of solar active regions (e.g. Fan 2009) and the onset of solar flares and coronal mass ejections (e.g. Török and Kliem 2003; Titov et al. 2003; Hood et al. 2016). A classic magnetohydrodynamic (MHD) model for magnetic flux tubes was first presented by Gold and Hoyle (1960), initially intended for use in the study of solar flares. The GH model is an infinite, straight, 1D and nonlinear force-free magnetic flux tube with constant 'twist' (Birn and Priest 2007). Mathematically, the GH magnetic field could be regarded as the cylindrical analogue of the Force-Free Harris sheet (Tassi et al. 2008), as the Bennett Pinch (1934) might be to the 'original' Harris Sheet.

It is typical to consider solar, space and astrophysical flux tubes within the framework of MHD (e.g. see Priest 2014). However, many of these plasmas can be weakly collisional or collisionless, with values of the collisional free path large against any fluid scale (Marsch 2006), making a description using collisionless kinetic theory necessary. In this chapter, it is our intention to study the GH flux tube model beyond the MHD description, since—apart from the very recent work in Vinogradov et al. (2016)—we see no attempt in the literature of a microscopic description of the GH field.

The work in Chaps. 2 and 3, as well as Alpers (1969), Harrison and Neukirch (2009a, b), Neukirch et al. (2009), Wilson and Neukirch (2011), Abraham-Shrauner (2013), Kolotkov et al. (2015), used methods like Channell's (Channell 1976) to tackle the VM inverse problem in Cartesian geometry. Channell described the extension of his work to cylindrical geometry as 'not possible in a straightforward manner.' As explained in Tasso and Throumoulopoulos (2014) (in which cylindrical coordinates are used to model a torus), this is due in part to the 'toroidicity' of the problem, i.e. the $1/r$ factor in the equations. As we shall see in this chapter, another potential complication is the need to allow—at least in principle—a non-zero charge density.

There has been significant recent work on VM equilibria that are consistent with nonlinear force-free (Harrison and Neukirch 2009a, b; Neukirch et al. 2009; Wilson and Neukirch 2011; Abraham-Shrauner 2013; Kolotkov et al. 2015; Allanson et al. 2015, 2016) and 'nearly force-free' (Artemyev 2011) magnetic fields in Cartesian geometry. VM equilibria for linear force-free fields have also been found in Sestero (1967), Bobrova and Syrovatskiĭ (1979), Bobrova et al. (2001). Therein, force-free refers to a magnetic field for which the associated current density is exactly parallel, which is the definition we shall also use,

$$\boldsymbol{j} \times \boldsymbol{B} = \frac{1}{\mu_0}(\nabla \times \boldsymbol{B}) \times \boldsymbol{B} = \boldsymbol{0}.$$

These works consider 1D collisionless current sheets, and so a natural question to consider is whether it is also possible to find self-consistent force-free (or nearly force-free) VM equilibria for other geometries, in particular cylindrical geometry. In this chapter we shall present particular VM equilibria for 1D magnetic fields which are nearly force-free in cylindrical geometry, i.e. flux tubes/ropes. These kinetic models and the the theory that follows are of potential applicability in the solar corona (e.g. see Wiegelmann and Sakurai 2012; Hood et al. 2016), Earth's magnetotail (e.g. see Kivelson and Khurana 1995; Khurana et al. 1995; Slavin et al. 2003; Yang et al. 2014) and magnetopause (e.g. Eastwood et al. 2016), planetary magnetospheres (e.g. DiBraccio et al. 2015), tokamak (e.g. Tasso and Throumoulopoulos 2007, 2014) and laboratory (e.g. Davidson 2001) plasmas.

5.2.1 Previous Work

Two of the archetypal field configurations in cylindrical geometry are the z-Pinch and the θ-pinch. The z-pinch has axial current and azimuthal magnetic field,

$$\boldsymbol{j} \times \boldsymbol{B} = \nabla p \iff \frac{d}{dr}\left(p + \frac{B_\theta^2}{2\mu_0}\right) + \frac{B_\theta^2}{\mu_0 r} = 0,$$

Freidberg (1987), a classical example of which is the Bennett Pinch

$$\tilde{B}_\theta = -\frac{\tilde{r}}{1+\tilde{r}^2}, \tag{5.1}$$

written in non-dimensional units, and for which a Vlasov equilibrium is well known (Bennett 1934; Harris 1962). In contrast, the θ-Pinch has azimuthal current and axial magnetic field,

$$\boldsymbol{j} \times \boldsymbol{B} = \nabla p \iff \frac{d}{dr}\left(p + \frac{B_z^2}{2\mu_0}\right) = 0.$$

Pinches that have both axial and azimuthal magnetic fields are known as screw or cylindrical pinches, e.g. see Freidberg (1987); Carlqvist (1988).

Consideration of 'Vlasov-fluid' models of z-Pinch equilibria was given in Channon and Coppins (2001), with Mahajan (1989) calculating z-Pinch equilibria and an extension with azimuthal ion-currents. Others have also constructed kinetic models of the θ-pinch, see Nicholson (1963); Batchelor and Davidson (1975) for examples.

In the same year as Pfirsch (1962), cylindrical kinetic equilibria with only azimuthal currents were studied in Komarov and Fadeev (1962). For examples of treatments of the stability of fluid and kinetic linear pinches, see Newcomb (1960), Pfirsch (1962), Davidson (2001) respectively.

Recently there have been studies on 'tokamak-like' VM equilibria with flows (Tasso and Throumoulopoulos 2007, 2014), starting from the VM equation in cylindrical geometry and working towards Grad-Shafranov equations for the vector potential. We also note two Vlasov equilibrium DFs in the literature that are close in style to the one that we shall present. The first is described in a brief paper (El-Nadi et al. 1976), with an equilibrium presented for a cylindrical pinch. However, their distribution describes a different magnetic field and the DF appears not to be positive over all phase space. The second DF is a very recent paper that actually describes a magnetic field much like the one that we discuss (Vinogradov et al. 2016). Their DF is designed to model 'ion-scale' flux tubes in the Earth's magnetosphere. Formally, their quasineutral model approaches a nonlinear force-free configuration in the limit of a vanishing electron to ion mass ratio. In their model, current is carried exclusively by electrons and the non-negativity of the DF depends on a suitable choice of microscopic parameters. Finally, we mention that in beam physics (e.g. see Morozov and Solov'ev 1961; Hammer and Rostoker 1970; Gratreau and Giupponi 1977; Uhm and Davidson 1985), much work on constructing cylindrical VM equilibria is done by looking for mono-energetic distributions with conserved angular momentum,

$$f_s = \delta(H_s - H_{0s})g(p_{\theta s}),$$

for H_{0s} a fixed energy, H_s and $p_{\theta s}$ the Hamiltonian and angular momentum respectively.

This chapter is structured as follows. In Sect. 5.3 we first review the theory of the equation of motion consistent with a collisionless DF in cylindrical geometry, and discuss the question of the possibility of 1D force-free equilibria. Then we introduce the magnetic field to be used. We note that whilst the work in this chapter is applied to a particular magnetic field from Sect. 5.3.6 onwards, the steps taken to calculate the equilibrium DF seem as though they could be adaptable to other cases. In Sect. 5.4 we present the form of the DF that gives the required macroscopic equilibrium, and proceed to 'fix' the parameters of the DF by explicitly solving Ampère's Law and Poisson's Equation. Note that whilst we choose to consider a two-species plasma of ions and electrons, we see no obvious reason preventing the work in this chapter being used to describe plasmas with a different composition. In Sect. 5.5 we present a preliminary analysis of the physical properties of the equilibrium. The analysis includes discussions on non-neutrality and the electric field; the equation of state and the plasma beta; the origin of individual terms in the equation of motion; plots of the DF; as well as particularly technical calculations in Sects. 5.4.1, 5.5.4.1 and 5.5.4.2. Section 5.4.1 contains the zeroth and first order moment calculations, used to find the number densities and bulk flows directly, and in turn the charge and current densities. Sections 5.5.4.1 and 5.5.4.2 contain the mathematical details of the existence and location of multiple maxima of the DF in velocity-space.

The work in this chapter does not present a generalised method for the VM inverse problem in cylindrical geometry, but instead some particular solutions for a specific given magnetic field. Other than any interesting theoretical advances, a possible application of the results of this study could be to implement the obtained model in kinetic (particle) numerical simulations.

5.3 General Theory

5.3.1 Vlasov Equation in Time-Independent Orthogonal Coordinates

A collisionless equilibrium is characterised by the 1-particle DF, f_s, a solution of the steady-state Vlasov Equation (e.g. see Schindler 2007). The Vlasov equation can be written (Santini and Tasso 1970) in index notation as

$$\frac{\partial f_s}{\partial t} + \frac{1}{\sqrt{g}}\frac{\partial}{\partial x^i}\left(\sqrt{g}\frac{dx^i}{dt}f_s\right) + \frac{\partial}{\partial v^i}\left(\frac{dv^i}{dt}f_s\right) = 0, \tag{5.2}$$

for $i \in \{1, 2, 3\}$; time-independent orthogonal coordinates given by $x^i \in (x^1, x^2, x^3)$; orthogonal and orthonormal basis vectors defined by e_i and $\hat{e}_i$ respectively; the diagonal metric tensor $g_{ij} = g_{ij}(x^1, x^2, x^3) = e_i e_j$, such that distances in configuration-space obey

$$ds^2 = g_{11}(dx^1)^2 + g_{22}(dx^2)^2 + g_{33}(dx^3)^2;$$

$g = \text{Det}[g_{ij}] = g_{11}g_{22}g_{33}$; velocities given by $\boldsymbol{v} = v^i\hat{e}_i = \sqrt{g_{ii}}\,dx^i/dt\,\hat{e}_i$; and the Einstein summation convention applied such that repeated indicies are summed over, i.e.

$$A_i B^i = A_1 B^1 + A_2 B^2 + A_3 B^3.$$

Superscript and subscript indices represent contra- and co-variant tensor components respectively, with the metric tensor able to raise or lower these indices, e.g.

$$x_j = g_{ij}x^i,$$

such that

$$\nabla = e_i\nabla^i = g_{ij}e^j\nabla^i = e^j\nabla_j (= e^i\nabla_i = \nabla),$$

and

$$\nabla_i = \frac{\partial}{\partial x^i}$$

(see e.g. Leonhardt and Philbin 2012; Landau and Lifshitz 2013 for good introductions to index notation).

Equation (5.2) can be re-written in vector notation (Santini and Tasso 1970) as

$$\frac{\partial f_s}{\partial t} + \boldsymbol{v} \cdot \nabla f + \frac{q_s}{m_s} [(\boldsymbol{E} + \boldsymbol{v} \times \boldsymbol{B}) - \boldsymbol{v} \times (\nabla \times \boldsymbol{v})] \cdot \frac{\partial f_s}{\partial \boldsymbol{v}} = 0, \tag{5.3}$$

for

$$\boldsymbol{v} = v^i \hat{e}_i, \quad \frac{\partial f_s}{\partial \boldsymbol{v}} = \hat{e}^i \frac{\partial f_s}{\partial v^i},$$

In Cartesian geometry, Eq. (5.3) reduces to a familiar form since the Cartesian basis vectors are position-independent, i.e.

$$\nabla \times \boldsymbol{v} = v^x \nabla \times \hat{e}_x + v^y \nabla \times \hat{e}_y + v^z \nabla \times \hat{e}_z = \boldsymbol{0}.$$

5.3.2 Vlasov Equation in Cylindrical Geometry

In cylindrical geometry $(x^1 = r, x^2 = \theta, x^3 = z)$, $(\hat{e}_r = e_r = \hat{r}, \hat{e}_\theta = \frac{1}{r} e_\theta = \hat{\theta}, \hat{e}_z = e_z = \hat{z})$, $\nabla \times \boldsymbol{v} = v^\theta \hat{r} \times \nabla \theta$, and Eq. (5.3) can be shown to reduce to

$$\frac{\partial f_s}{\partial t} + \boldsymbol{v} \cdot \nabla f_s + \frac{q_s}{m_s} (\boldsymbol{E} + \boldsymbol{v} \times \boldsymbol{B}) \cdot \frac{\partial f_s}{\partial \boldsymbol{v}} + \left[\frac{v_\theta^2}{r} \frac{\partial f_s}{\partial v_r} - \frac{v_r v_\theta}{r} \frac{\partial f_s}{\partial v_\theta} \right] = 0, \tag{5.4}$$

e.g. see Komarov and Fadeev (1962), Santini and Tasso (1970) and Tasso and Throumoulopoulos (2007). Note that the gradient operator in cylindrical coordinates is given by

$$\nabla = \hat{r} \frac{\partial}{\partial r} + \hat{\theta} \frac{1}{r} \frac{\partial}{\partial \theta} + \hat{z} \frac{\partial}{\partial z},$$

such that the matrix representation of the metric tensor, $\underline{\underline{g}} = \text{Mat}[g_{ij}]$, is given by

$$\underline{\underline{g}} = \begin{pmatrix} 1 & 0 & 0 \\ 0 & r^2 & 0 \\ 0 & 0 & 1 \end{pmatrix}.$$

The 'fluid' equation of motion of a particular species s is found by taking first-order velocity moments of the Vlasov equation. For the purposes of completeness and future reference the full first order moment-taking calculation is performed in Sect. 5.3.3, since it is not easily found in the literature, to our knowledge. The result is that for an arbitrary DF that only depends spatially on r, the equation of motion can almost be written in a familiar form, as compared to the equation in Cartesian geometry (Mynick et al. 1979; Greene 1993; Schindler 2007), but with some 'additional' terms. This is to be expected, given the form of Eq. (5.4).

5.3.3 Equation of Motion in Cylindrical Geometry

It will be useful to-rewrite the Vlasov equation from Eq. (5.4) in index notation, in order to take the velocity moments. As such, the Vlasov equation can be written according to

$$\frac{\partial f_s}{\partial t} + v^i \nabla_i f_s + \frac{q_s}{m_s}\left(E_i + \varepsilon_{ijk} v^j B^k\right) \nabla_{v_i} f_s + \left[\frac{(v^\theta)^2}{r} \nabla_{v^r} f_s - \frac{v^r v^\theta}{r} \nabla_{v^\theta} f_s\right] = 0. \tag{5.5}$$

The totally antisymmetric unit tensor of rank 3 (the Levi-Civita tensor) is ε_{ijk}, and it takes the value 0 when any of its indices are repeated (e.g. $\varepsilon_{131} = 0$), $+\sqrt{g}$ for an 'ordered triplet' (e.g. $\varepsilon_{231} = \sqrt{g}$), and $-\sqrt{g}$ for a 'disordered triplet' (e.g. $\varepsilon_{213} = -\sqrt{g}$). The first moment of the Vlasov equation (Eq. 5.5), and multiplied by m_s, gives

$$\begin{aligned} m_s \int \Bigg\{ & \underbrace{v_i \frac{\partial f_s}{\partial t}}_{\text{A}} + \underbrace{v_i v_j \nabla^j f_s}_{\text{B}} + \frac{q_s}{m_s}\Bigg(\underbrace{v_i E_j \nabla_{v_j} f_s}_{\text{C}} + \underbrace{v_i \epsilon_{jkl} v^k B^l \nabla_{v_j} f_s}_{\text{D}} \Bigg) \\ & + \underbrace{v_i \frac{(v^\theta)^2}{r} \nabla_{v^r} f_s}_{\text{E}} - \underbrace{v_i \frac{v^r v^\theta}{r} \nabla_{v^\theta} f_s}_{\text{F}} \Bigg\} d^3 v = 0, \end{aligned} \tag{5.6}$$

with the triple integral written in shorthand by

$$\int d^3 v := \int_{-\infty}^{\infty} \int_{-\infty}^{\infty} \int_{-\infty}^{\infty} dv_r dv_\theta dv_z.$$

The first term, 'A', gives $\partial/\partial t(\rho_s V_{js})$. Next, we notice that the spatial derivative in 'B' can be taken outside of the integral. Then, if we write $v_i = V_{is} + w_{js}$, we see that by Leibniz' rule, for ∇ a derivative

$$\nabla \langle v_i v_j \rangle = \nabla (V_{is} V_{js}) + \nabla \langle w_{is} w_{js} \rangle + \nabla (V_{is} \langle w_{js} \rangle) + \nabla (V_{js} \langle w_{is} \rangle),$$

with the angle brackets denoting an integral over velocity space (by definition $\langle w_i \rangle = 0$). As a result, 'B' becomes

$$\nabla^j (\rho_s V_{js} V_{is}) + \nabla^j P_{ij}.$$

We shall integrate terms 'C–F' by parts and neglect surface terms, i.e. we assume that

$$\lim_{|\boldsymbol{v}|\to\infty} \mathcal{G}(\boldsymbol{x}, \boldsymbol{v}, t) f_s(\boldsymbol{x}, \boldsymbol{v}, t) = 0,$$

for $\mathcal{G}$ representing the different variables multiplying the DF in terms 'C-F'. As a result 'C' and 'D' become $-\sigma_s E_i$ and $-\sigma_s \epsilon_{ijk} V_s^j B^k$ respectively. If again, we rewrite $v_i = V_{is} + w_{is}$, and use Leibniz' rule, 'E' becomes

$$-\frac{\delta_{ir}}{r}\rho_s V_{\theta s}^2 - \frac{\delta_{ir}}{r} P_{\theta\theta,s},$$

with δ_{ij} the Kronecker delta. Similarly, 'F' becomes

$$\frac{1}{r}\pi_{ir,s} + \frac{\delta_{i\theta}}{r}\pi_{r\theta,s},$$

for $\pi_{ij,s} = m_s \int v_i v_j f_s d^3v$. Putting this all together gives

$$\begin{aligned} &\rho_s \frac{\partial V_{is}}{\partial t} + \nabla^j P_{ij,s} + \nabla^j(\rho_s V_{js} V_{is}) - \sigma_s E_i - \sigma_s \epsilon_{ijk} V_s^j B^k \\ &-\rho_s (V_{\theta s})^2 \frac{\delta_{ir}}{r} - \frac{\delta_{ir}}{r} P_{\theta\theta,s} + \frac{1}{r}\pi_{ir,s} + \frac{\delta_{i\theta}}{r}\pi_{r\theta,s} = 0. \end{aligned} \tag{5.7}$$

Taking the r-component, in equilibrium ($\partial/\partial t = 0$), assuming a 1D configuration with only radial dependence ($\partial/\partial\theta = \partial/\partial z = 0$), letting f_s be an even function of v_r ($V_{rs} = P_{r\theta} = P_{zr} = 0$), and noticing that $\pi_{rr,s} = \rho_s V_{rs}^2 + P_{rr,s} = P_{rr,s}$ gives

$$\frac{\partial P_{rr,s}}{\partial r} + \frac{1}{r}(P_{rr,s} - P_{\theta\theta,s}) = \sigma_s(\boldsymbol{E} + \boldsymbol{V}_s \times \boldsymbol{B})_r + \rho_s \frac{V_{\theta s}^2}{r}.$$

We now consider the general expression for the r component of the divergence of a rank-2 tensor in cylindrical coordinates (Huba 2013)

$$(\nabla \cdot \boldsymbol{T})_r = \frac{1}{r}\frac{\partial}{\partial r}(rT_{rr}) + \frac{1}{r}\frac{\partial T_{\theta r}}{\partial \theta} + \frac{\partial T_{zr}}{\partial z} - \frac{T_{\theta\theta}}{r}. \tag{5.8}$$

Since the $P_{r\theta}$ and P_{zr} terms of the pressure tensor are zero, this becomes

$$(\nabla \cdot \boldsymbol{P})_r = \frac{1}{r}\frac{\partial}{\partial r}(rP_{rr}) - \frac{P_{\theta\theta}}{r}, \tag{5.9}$$

and so force balance for species s is maintained—in equilibrium ($\partial/\partial t = 0$), assuming a 1D configuration with only radial dependence ($\partial/\partial\theta = \partial/\partial z = 0$), and letting f_s be an even function of the radial velocity v_r—according to

$$(\nabla \cdot \boldsymbol{P}_s)_r = (\boldsymbol{j}_s \times \boldsymbol{B})_r + \sigma_s E_r + \frac{\rho_s}{r} V_{\theta s}^2. \tag{5.10}$$

Equation (5.10) can be summed over species to give

$$(\nabla \cdot \boldsymbol{P})_r + \mathcal{F}_{\mathrm{c}} = (\boldsymbol{j} \times \boldsymbol{B})_r + \sigma \boldsymbol{E}, \tag{5.11}$$

where

$$\mathcal{F}_{\mathrm{c}} = \sum_s \mathcal{F}_{\mathrm{c},s} = -\frac{1}{r}\left(\rho_i V_{\theta i}^2 + \rho_e V_{\theta e}^2\right)\hat{\boldsymbol{e}}_r$$

is the force density associated with the rotating bulk flows of the ions and electrons, and is in fact a centripetal force. Equation (5.11) is a cylindrical analogue of the force balance equation in Cartesian geometry (e.g. see Mynick et al. 1979). However, in the cylindrical case there are extra terms due to centripetal forces. Note that in a non-inertial frame that is co-moving with the respective species bulk flows, the species s will also feel a fictitious force equal to $-\mathcal{F}_{\mathrm{c},s}$ (as well as any other forces), and this is known as the centrifugal force.

From the point of view of a particular magnetic field $\boldsymbol{B}$ (which is the point we take by specifying a particular macroscopic equilibrium), we see that equilibrium is maintained by a combination of density/pressure variations as in the case of Cartesian geometry, but with additional contributions from centripetal forces and as an inevitable result of the resultant charge separation, an electric field. This effect is represented in Fig. 5.1, with Fig. 5.1a depicting the case for $E_r < 0$, such that $-\mathcal{F}_{ci} > -\mathcal{F}_{ce}$. Whereas Fig. 5.1b depicts the case for $E_r > 0$, such that $-\mathcal{F}_{ce} > -\mathcal{F}_{ci}$. This demonstrates that 'sourcing' an exactly force-free macroscopic equilibrium with an equilibrium DF in a 1D cylindrical geometry is inherently a more difficult task than in the Cartesian case. The presence of 'extra' centripetal forces, and almost inevitably forces associated with charge separation, raises the question of whether exactly force-free ($\boldsymbol{j} \times \boldsymbol{B} = \boldsymbol{0}$) equilibria are possible at all in this geometry.

Before proceeding, we comment that given certain macroscopic constraints on the electromagnetic fields or fluid quantities—such as the force-free condition, or a specific given magnetic field (for example)—it is not *a priori* known how to calculate a self-consistent Vlasov equilibrium, or if one even exists within the framework of the assumptions made. Hence one has to proceed more or less on a case by case basis, with the intention of achieving consistency with the required macroscopic conditions, upon taking moments of the DF.

5.3.4 The Gold-Hoyle (GH) Magnetic Field

The GH magnetic field (Gold and Hoyle 1960) is a 1D ($\partial/\partial\theta = \partial/\partial z = 0$), nonlinear force-free ($\nabla \times \boldsymbol{B} = \alpha(r)\boldsymbol{B}$) and uniformly twisted flux-tube model, with

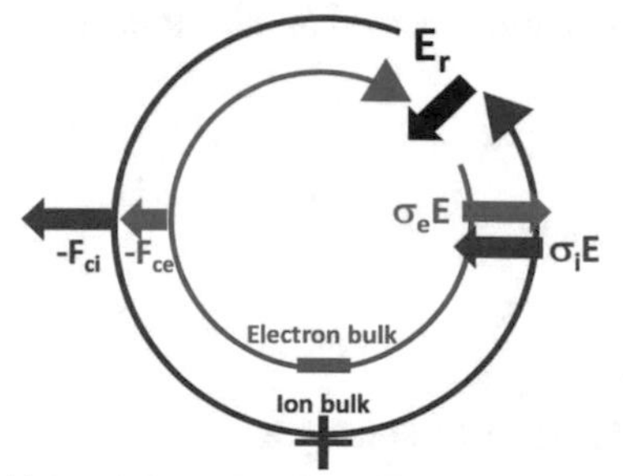

(a) Force balance with $E_r < 0$ and $-\mathcal{F}_{ci} > -\mathcal{F}_{ce}$

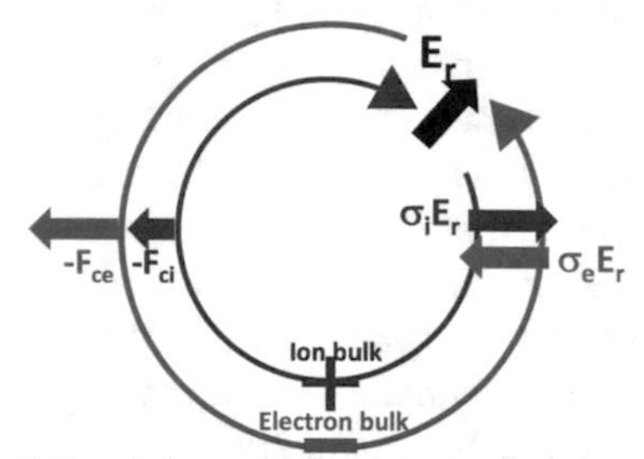

(b) Force balance with $E_r > 0$ and $-\mathcal{F}_{ce} > -\mathcal{F}_{ci}$

Fig. 5.1 A schematic representation of how, in force balance, the electric field, E_r exists in order to balance the 'charge separation' effect caused by the forces associated with the ion and electron rotational bulk flows, $\mathcal{F}_{ci}$ and $\mathcal{F}_{ce}$ respectively. Figure 5.1a depicts the case for $E_r < 0$, such that $-\mathcal{F}_{ci} > -\mathcal{F}_{ce}$, whilst Fig. 5.1b depicts the case for $E_r > 0$, such that $-\mathcal{F}_{ce} > -\mathcal{F}_{ci}$

$$\begin{aligned}
\boldsymbol{A}_{GH}(\tilde{r}) &= \frac{B_0}{2\tau}\left(0, \frac{1}{\tilde{r}}\ln\left(1+\tilde{r}^2\right), -\ln\left(1+\tilde{r}^2\right)\right), \\
\boldsymbol{B}_{GH}(\tilde{r}) &= B_0\left(0, \frac{\tilde{r}}{1+\tilde{r}^2}, \frac{1}{1+\tilde{r}^2}\right), \\
\boldsymbol{j}_{GH}(\tilde{r}) &= 2\frac{\tau B_0}{\mu_0}\left(0, \frac{\tilde{r}}{(1+\tilde{r}^2)^2}, \frac{1}{(1+\tilde{r}^2)^2}\right), \\
\boldsymbol{j}_{GH}(\boldsymbol{A}, \tilde{r}) &= 2\frac{\tau B_0}{\mu_0}\left(0, \tilde{r}e^{-\frac{4\tau}{B_0}\tilde{r}A_\theta}, e^{\frac{4\tau}{B_0}A_z}\right),
\end{aligned} \tag{5.12}$$

The constant τ has units of inverse length, and we use $1/\tau$ to represent the characteristic length scale of the system ($\tilde{r} = \tau r$). The parameter B_0 gives the magnitude of the magnetic field at $\tilde{r} = 0$. Note that the representation of $\boldsymbol{j}_{GH}(\boldsymbol{A})$ chosen in Eq. (5.12) is representative and non-unique. In fact there are other possible representations, that include 'mixtures' of A_θ and A_z in each component of the current density.

Furthermore, τ is a direct measure of the 'twist' of the embedded flux tube (see Birn and Priest 2007), with the number of turns per unit length (in z) along a field line given by $\tau/(2\pi)$ (Gold and Hoyle 1960). A diagram representing the qualitative interior structure of such a flux tube is given in Fig. 5.2, and reproduced from Russell and Elphic (1979) (their magnetic field was in fact not quite uniformly twisted, but close enough that the diagram still serves a purpose). The most important feature to note is how the B_z component of the field dominates at small radii, whereas the B_θ component dominates for larger radii. This characteristic ensures that you travel the

Fig. 5.2 The interior structure of a flux tube, from Russell and Elphic (1979), and similar to the GH model. Image Copyright: Nature Publishing Group. Reprinted by permission from Macmillan Publishers Ltd: *Nature* **279** (June 1979), pp. 616–618. copyright (1979)

same distance in z, for each 2π revolution, regardless of how far from the central axis you are ($d\theta/dz =$ const.). The force-free parameter for the magnetic field is

$$\alpha(r) = \frac{\nabla \times \boldsymbol{B} \cdot \boldsymbol{B}}{|\boldsymbol{B}|^2} = \frac{2\tau}{1+\tilde{r}^2}.$$

Should one wish to consider the GH field in an MHD context ($\nabla p = \boldsymbol{j} \times \boldsymbol{B} = \boldsymbol{0}$) then the scalar pressure $p =$ const.. This is seen by considering the 1D force-balance equation (Freidberg 1987),

$$\boldsymbol{j} \times \boldsymbol{B} = \nabla p \iff \frac{d}{dr}\left(p + \frac{B_\theta^2}{2\mu_0} + \frac{B_z^2}{2\mu_0}\right) + \frac{B_\theta^2}{\mu_0 r} = 0,$$

for the GH field.

5.3.5 *Methods for Calculating an Equilibrium DF*

In Channell (1976), Harrison and Neukirch (2009a) for example, a method used to calculate a DF, given a prescribed 1D magnetic field was Inverse Fourier Transforms (IFT). This method was also discussed in Sect. 1.3.5.5. A DF of the form

$$f_s \propto e^{-\beta_s H_s} g_s(p_{xs}, p_{ys}), \tag{5.13}$$

was used, with H_s, p_{xs} and p_{ys} the conserved particle Hamiltonian and canonical momenta in the x and y directions, and g_s an unknown function, to be determined. Since our problem is one of a 1D equilibrium with variation in the radial direction, the three constants of motion are the Hamiltonian, and the canonical momenta in the θ and z directions:

$$\begin{aligned} H_s &= \frac{m_s}{2}\left(v_r^2 + v_\theta^2 + v_z^2\right) + q_s\phi, \\ p_{\theta s} &= r\left(m_s v_\theta + q_s A_\theta\right), \quad p_{zs} = m_s v_z + q_s A_z. \end{aligned} \tag{5.14}$$

One can try to calculate an equilibrium distribution for the GH force-free flux tube without a background field by a similar method, assuming a DF of the form

$$f_s \propto e^{-\beta_s H_s} g_s(p_{\theta s}, p_{zs}). \tag{5.15}$$

By exploiting the convolution in the definition of the current density,

$$\begin{aligned} \boldsymbol{j}(\boldsymbol{A}, r) &= \sum_s q_s \int \boldsymbol{v}\, f_s(H_s, p_{\theta s}, p_{zs})\, d^3v, \\ &= r \sum_s \frac{q_s}{m_s^4} \int (\mathfrak{p}_s - q_s \boldsymbol{A})\, f_s(H_s, r\mathfrak{p}_{\theta s}, \mathfrak{p}_{zs})\, d^3\mathfrak{p}_s, \end{aligned}$$

Ampère's law can be solved formally by IFT (*cf.* Harrison and Neukirch 2009a and Sect. 1.3.5.5), or informally by 'inspection' (*cf.* Neukirch et al. 2009), with the quantity $\mathfrak{p}_s$ defined by

$$\mathfrak{p}_{rs} = p_{rs}, \quad \mathfrak{p}_{\theta s} = \frac{p_{\theta s}}{r}, \quad \mathfrak{p}_{zs} = p_{zs}.$$

Notice how when written in this integral form, $\boldsymbol{j}$ is not only a function of $\boldsymbol{A}$, but—in contrast with the Cartesian case—also of the relevant spatial co-ordinate, r.

5.3.5.1 Problems with Equilibrium DFs for the GH Field

We shall now reproduce the calculations, representatively, for the j_θ case. These calculations are representative in that the choice of expression for the current density as a function of the vector potential is non-unique, as indicated previously. However, this calculation should demonstrate the inherent obstacle in calculating a Vlasov equilibrium DF for the GH field.

The definition of the current density, along with the ansatz of Eq. (5.15) gives

$$j_\theta = r \sum_s \frac{q_s}{m_s^4} \frac{n_{0s}}{(\sqrt{2\pi} v_{\mathrm{th},s})^3} e^{-\beta_s q_s \phi} \int (\mathfrak{p}_{\theta s} - q_s A_\theta) e^{-(\mathfrak{p}_s - q_s \mathbf{A})^2/(2m_s^2 v_{\mathrm{th},s}^2)} g_s(r\mathfrak{p}_{\theta s}, \mathfrak{p}_{zs}) d^3\mathfrak{p}_s.$$

If we now take a representative (i.e. one possible) expression for the current density, chosen as a more 'general' form than that in Eq. (5.12),

$$j_\theta = c_1 \frac{\tau^2 B_0 r}{\mu_0} \exp\left(\frac{c_2 \tau^2 r A_\theta}{B_0} + \frac{c_3 \tau A_z}{B_0} \right),$$

for c_1, c_2 and c_3 constants, and re-write $\mathfrak{p}_{\theta s} = p_{\theta s}/r$, then we obtain

$$c_1 \frac{\tau^2 B_0 r}{\mu_0} \exp\left(\frac{c_2 \tau^2 r A_\theta}{B_0} + \frac{c_3 \tau A_z}{B_0}\right) = \sum_s \frac{q_s}{m_s^3} \frac{n_{0s}}{(\sqrt{2\pi} v_{\text{th},s})^2} e^{-\beta_s q_s \phi} \times$$

$$\int (p_{\theta s} - q_s r A_\theta) e^{-(p_{\theta s} - q_s r A_\theta)^2/(2m_s^2 v_{\text{th},s}^2 r^2) - (p_{zs} - q_s A_z)^2/(2m_s^2 v_{\text{th},s}^2)} g_s(p_{\theta s}, p_{zs}) dp_{\theta s} dp_{zs}.$$

In the case of zero scalar potential, the result of the calculation is to give a g_s function (and hence a DF) that is not a solution of the Vlasov equation as it is not a function of the constants of motion only. In essence, an additional "exp($-r^2$)" factor would be required in the DF to counter "exp($+r^2$)" terms that manifest by completing the square in the integration. That is to say that the 'solution' would be of the form

$$g_s(p_{\theta s}, p_{zs}) = g_0 \exp\left(-\frac{\omega_s^2}{2\tau^2 v_{\text{th},s}^2} \delta_s^2 \tau^2 r^2\right) \exp\left(\frac{\omega_s}{\tau v_{\text{th},s}} \frac{\tau^2 p_{\theta s}}{q_s B_0} + \frac{V}{v_{\text{th},s}} \frac{\tau p_{zs}}{q_s B_0}\right),$$

and hence the DF can be written as

$$f_s \propto g_0 \exp\left(-\frac{\omega_s^2}{2\tau^2 v_{\text{th},s}^2} \delta_s^2 \tau^2 r^2\right) e^{-\beta_s H_s} \exp\left(\frac{\omega_s}{\tau v_{\text{th},s}} \frac{\tau^2 p_{\theta s}}{q_s B_0} + \frac{V}{v_{\text{th},s}} \frac{\tau p_{zs}}{q_s B_0}\right), \quad (5.16)$$

for some g_0, ω_s and V related to c_1, c_2 and c_3 respectively. The ratio of the thermal Larmor radius, $r_L = m_s v_{\text{th},s}/(e|B|)$ (for $e = |q_s|$) to the macroscopic length scale of the system $L(= 1/\tau)$, is given by

$$\delta_s(r) = \frac{r_L}{L} = \frac{m_s v_{\text{th},s} \tau}{e B(r)},$$

typically known as the 'magnetisation parameter' (Fitzpatrick 2014) (see Table 5.1 for a concise list of the micro and macroscopic parameters of the equilibrium). Note that in our system, the magnitude of the magnetic field and hence δ_s itself is spatially variable. For the purposes of the calculations in this chapter however, we set

$$\frac{m_s v_{\text{th},s} \tau}{e B_0} = \delta_s = \text{const.}$$

as a characteristic value.

The DF in Eq. (5.16) is not a solution of the Vlasov equation, but would approximate one in the limit

$$\frac{\omega_s}{\tau v_{\text{th},s}} \delta_s = \frac{\omega_s}{q_s B_0/m_s} \to 0,$$

i.e. the vanishing ratio of the bulk angular frequency to the gyrofrequency of the individual particles (*cf.* Vinogradov et al. 2016 and more on this later). It is now apparent that the physical cause for the extra "exp($+r^2$)" term here would appear to

Table 5.1 The fundamental parameters of the equilibrium. The s subscript refers to particles of species s

Macroscopic parameter	Meaning
B_0	Characteristic magnetic field strength
τ	Measure of the twist of flux tube
k	Strength of the background field
$\gamma_1 \neq 0, 1, 0 < \gamma_2 < 1$	Gauge for scalar potential
U_{zs}, V_{zs}	Bulk rectilinear flows
ω_s	Bulk angular frequency
Microscopic parameter	Meaning
m_s	Mass of particle
q_s, e	Charge, magnitude of charge
$\beta_s = 1/(k_B T_s)$	Thermal beta
$v_{\mathrm{th},s}$	Thermal velocity
$\delta_s(r), \delta_s$	Magnetisation parameters
n_{0s}	Normalisation of particle number

be the forces associated with the rotational bulk flow, since the term is non-negligible when ω_s is of a sufficient magnitude.

If one assumes a non-zero scalar potential, then the above considerations would seem to imply that

$$-\beta_i q_i \phi = -\beta_e q_e \phi = -\frac{\omega_s^2}{2\tau^2 v_{\mathrm{th},s}^2}\delta_s^2 \tau^2 r^2,$$

for there to be an exact Vlasov solution. This equation cannot be satisfied. The physical cause seems to be that, in the case of force-free fields, one would require a 'different' electrostatic potential to balance the forces for the ions and electrons, which is of course nonsensical. Thus, our investigation seems to suggest that it is not possible to calculate a DF of the form of Eq. (5.15) for the exact GH field.

5.3.6 *GH Flux Tube Plus Background Field (GH+B)*

To make progress, we introduce a background field in the negative z direction. The mathematical motivation for this change is to balance the 'exp(r^2) problem'. Physically, it seems that the background field introduces an extra term (whose sign depends on species) into the force-balance, to allow for both the ion and electrons to be in force balance simultaneously, given one unique expression for the scalar potential.

The vector potential, magnetic field and current density used are as follows:

Fig. 5.3 The twist (normalised by $\tau/(2\pi)$) of the GH+B field for three values of k. Figure 5.3a shows the twist for $k < 1/2$, and as such there are both negative and positive twists, due to the field reversal. Figure 5.3b and c both show negative twist, since there is no magnetic field reversal

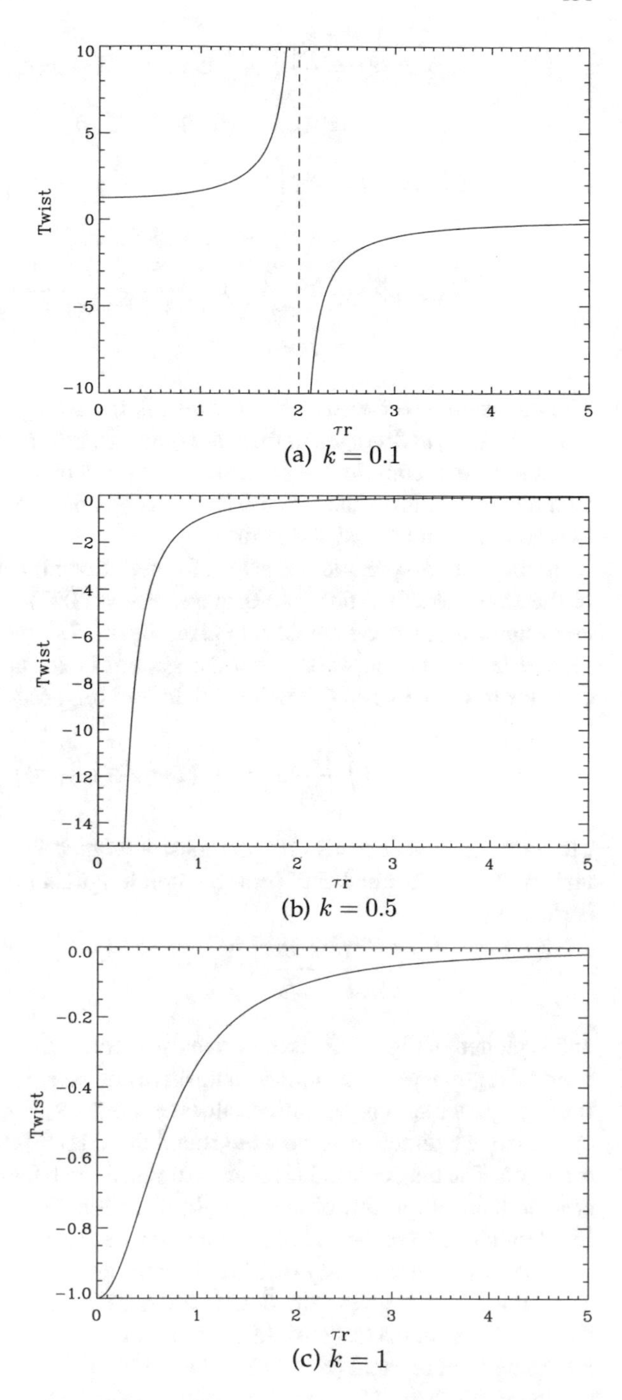

$$\begin{aligned}\boldsymbol{A}_{GH+B}(\tilde{r}) &= \frac{B_0}{2\tau}\left(0, \frac{1}{\tilde{r}}\ln\left(1+\tilde{r}^2\right) - 2k\tilde{r}, -\ln\left(1+\tilde{r}^2\right)\right), \\ &= \boldsymbol{A}_{GH} - \left(0, B_0 k\tau^{-1}\tilde{r}, 0\right). \end{aligned} \tag{5.17}$$

$$\begin{aligned}\boldsymbol{B}_{GH+B}(\tilde{r}) &= B_0\left(0, \frac{\tilde{r}}{1+\tilde{r}^2}, \frac{1}{1+\tilde{r}^2} - 2k\right), \\ &= \boldsymbol{B}_{GH} - (0, 0, 2kB_0). \end{aligned} \tag{5.18}$$

$$\begin{aligned}\boldsymbol{j}_{GH+B}(\tilde{r}) &= 2\frac{\tau B_0}{\mu_0}\left(0, \frac{\tilde{r}}{(1+\tilde{r}^2)^2}, \frac{1}{(1+\tilde{r}^2)^2}\right), \\ &= \boldsymbol{j}_{GH}. \end{aligned} \tag{5.19}$$

The dimensionless constant $k > 0$ controls the strength of the background field in the z direction, and as a result there are now two different interpretations to be made. We could either consider the system as a GH flux tube of uniform twist embedded in an untwisted uniform background field, or consider the whole GH+B magnetic field as a non-uniformly twisted flux tube.

In the first interpretation, τ is (as aforementioned) a direct measure of the 'twist' of the embedded flux tube (see Birn and Priest 2007), with the number of turns per unit length (in z) along a field line given by $\tau/(2\pi)$ (Gold and Hoyle 1960). In the second interpretation, we see that the system is not uniformly twisted, with the z distance traversed when following a field line (e.g. Marsh 1996), given by

$$\int \frac{rB_z}{B_\theta}d\theta = \frac{1}{\tau}\left(1 - 2k(1+\tilde{r}^2)\right)\int d\theta.$$

The fact that this depends on r demonstrates that the system as a whole has non-uniform twist. The number of turns per unit length in z of the GH+B field: the 'twist' is given by

$$\left(\int_{\theta=0}^{\theta=2\pi}\frac{rB_z}{B_\theta}d\theta\right)^{-1} = \frac{\tau}{2\pi}\left(1 - 2k(1+\tilde{r}^2)\right)^{-1},$$

and is plotted in Fig. 5.3 for three values of k. Since $k < 1/2$ corresponds to the field-reversal regime, we see a mixture of positive and negative twists (Fig. 5.3a). However, for $k \geq 1/2$ we see only negative values of the twist (Fig. 5.3b and c), i.e. we travel in the negative z direction as we wind round the GH+B flux tube in the anti-clockwise direction. The magnetic field is plotted in Fig. 5.4a–b for two values of k. The $k = 0.3$ case contains a reversal of the $\tilde{B}_z$ field direction and as such is akin to a Reversed Field Pinch (e.g. see Escande 2015 for a laboratory interpretation): this configuration may be of use in the study of astrophysical jets, see Li et al. (2006) for example. The value $k = 1/2$ gives zero $\tilde{B}_z$ at $\tilde{r} = 0$, and as such is the value that distinguishes the two different classes of field configuration, namely unidirectional ($k \geq 1/2$) or including field reversal ($k < 1/2$). The value of $\tilde{r}$ for which the $\tilde{B}_z$ field reverses is plotted in Fig. 5.4c. The magnitude of the GH+B magnetic field is plotted in Fig. 5.5 for three values of k. For all values of k, $|\tilde{\boldsymbol{B}}| \rightarrow 2k$ for large $\tilde{r}$, i.e. to a potential field.

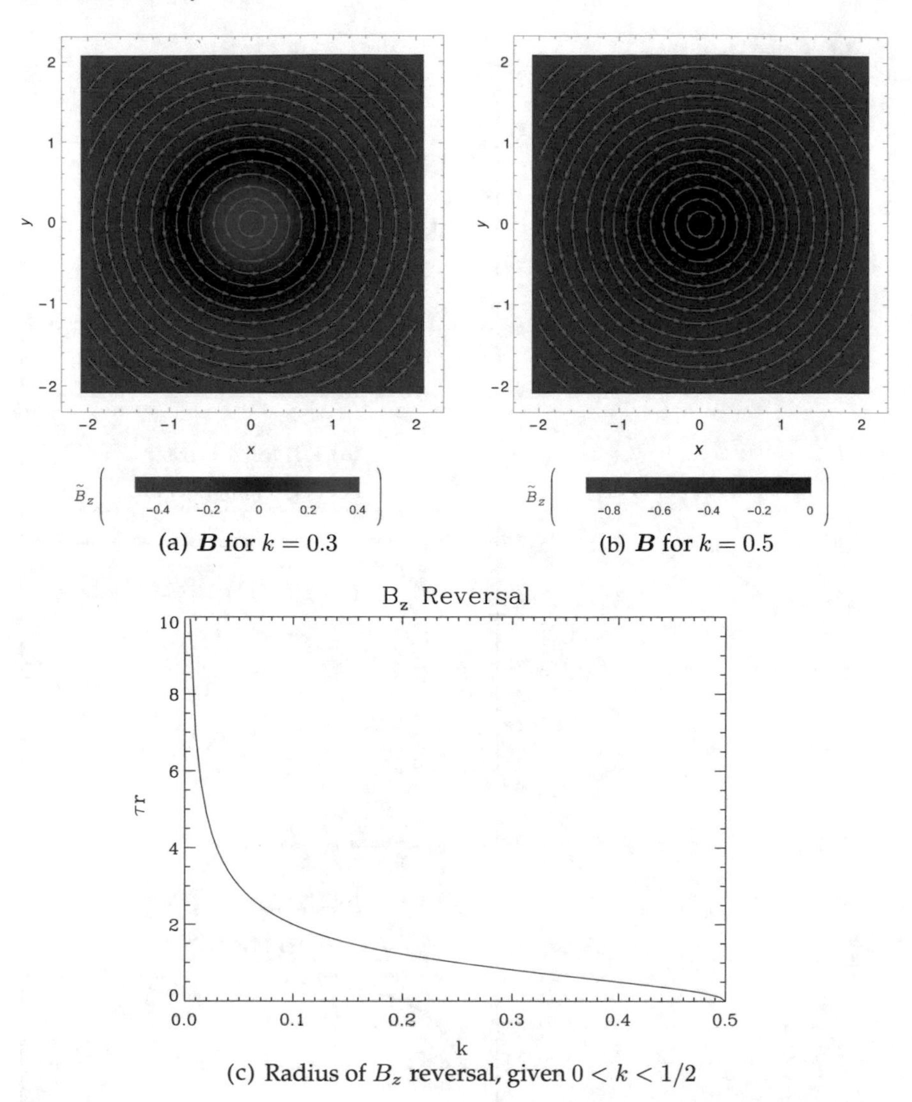

(a) $\boldsymbol{B}$ for $k = 0.3$

(b) $\boldsymbol{B}$ for $k = 0.5$

(c) Radius of B_z reversal, given $0 < k < 1/2$

Fig. 5.4 Figure 5.4a and b show the GH+B magnetic field in the xy plane, for two values of k. The curved arrows indicate the direction of the $\tilde{B}_\theta$ components, whilst the blue-black-red shading denotes the magnitude and direction of the $\tilde{B}_z$ component. The $k = 0.3$ case contains a reversal of the $\tilde{B}_z$ field direction and as such is a Reversed Field Pinch whilst $k = 0.5$ gives zero $\tilde{B}_z$ at $\tilde{r} = 0$. Figure 5.4c shows the radius at which $\tilde{B}_z$ changes its direction, for $0 < k < 1/2$. $\tilde{B}_z$ does not reverse for $k \geq 1/2$

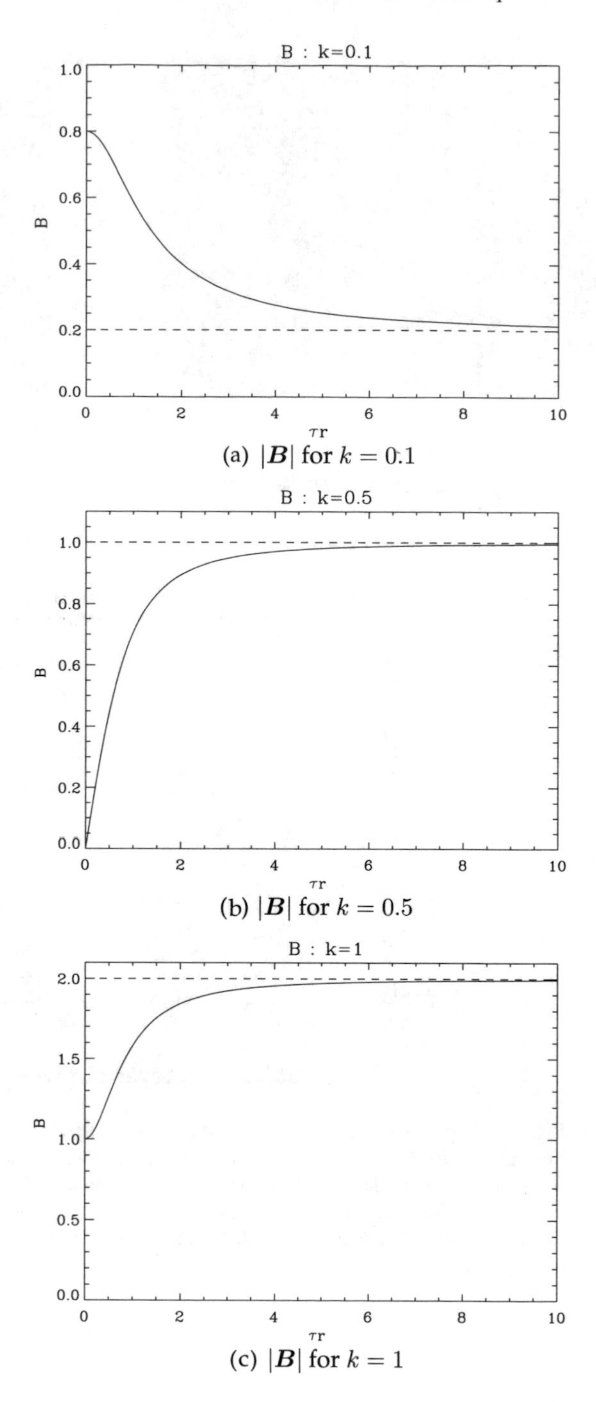

(a) $|\boldsymbol{B}|$ for $k = 0.1$

(b) $|\boldsymbol{B}|$ for $k = 0.5$

(c) $|\boldsymbol{B}|$ for $k = 1$

Fig. 5.5 Figure 5.5a–c show the magnitude of the GH+B magnetic field for $k = 0.1, 0.5$ and $k = 1$ respectively, normalised by B_0. For $k < 0.5$, $|\tilde{\boldsymbol{B}}| \to 2k$ from above, whereas for $k \geq 1/2$, $|\tilde{\boldsymbol{B}}| \to 2k$ from below

We also note here that flux tubes embedded in an axially directed background field have recently been observed during reconnection events in the Earth's magnetotail, by the Cluster spacecraft (e.g. Borg et al. 2012), and that recent numerical modelling of 'magnetohydrodynamic (MHD) avalanches' in the low-beta solar corona has used multiple flux ropes embedded in a uniform background magnetic field (Hood et al. 2016). The magnetic field model used (Hood et al. 2009) is similar to the model in this chapter, as it is force-free and 1D.

The primary task of this chapter is to calculate self-consistent collisionless equilibrium DFs for the GH+B field. This problem essentially reduces to solving Ampère's Law such that Eq. (5.4) is satisfied. We assume nothing about the electric field however, and in fact use that degree of freedom to solve Ampère's Law. The resultant form of the scalar potential is then substituted into Poisson's equation, to establish the final relationships between the microscopic and macroscopic parameters of the equilibrium.

5.4 The Equilibrium DF

Although the IFT method did not yield a self-consistent equilibrium DF for the GH field without a background field, the outcome of the calculation can still be used as an indication of possible forms for the DF for the GH+B field. Using trial and error we arrived at the DF

$$f_s = \frac{n_{0s}}{(\sqrt{2\pi}v_{\mathrm{th},s})^3}\left[e^{-(\tilde{H}_s-\tilde{\omega}_s\tilde{p}_{\theta s}-\tilde{U}_{zs}\tilde{p}_{zs})}+C_s e^{-(\tilde{H}_s-\tilde{V}_{zs}\tilde{p}_{zs})}\right], \tag{5.20}$$

which is a superposition of two terms that are consistent macroscopically with a 'Rigid-Rotor' (Davidson 2001). A Rigid-Rotor is microscopically described by a DF of the form $F(\mathcal{H}-\omega p_\theta - Vp_z)$. Each $F(H-\omega p_\theta - Vp_z)$ term corresponds to an average macroscopic motion of rigid rotation with angular frequency ω, and rectilinear motion with velocity V (with $\omega = 0$ in the second term of the DF in Eq. 5.20). This can be shown in a manner similar to that shown in Sect. 1.3.2.2.

The dimensionless constants $\tilde{\omega}_s$, $\tilde{U}_{zs}$, $\tilde{V}_{zs}$ and C_s are yet to be determined, with $C_s > 0$ for positivity of the distribution (see Table 5.2 for a concise list of the dimensionless quantities used in this chapter).

5.4.1 *Moments of the DF*

In order to satisfy Maxwell's equations, we shall require the charge and current densities. Hence we will require the zeroth- and first-order moments of the DF in Eq. (5.20), and these calculations follow. See Table 5.2 for a clarification of all dimensionless quantities denoted by a tilde, $\tilde{}$.

Table 5.2 Dimensionless form of some important variables. The s subscript refers to particles of species s

Variable	Dimensionless form
Particle Hamiltonian	$\tilde{H}_s = \beta_s H_s$
Particle angular momentum	$\tau p_{\theta s} = m_s v_{\text{th},s} \tilde{p}_{\theta s}$
Particle z-Momentum	$p_{zs} = m_s v_{\text{th},s} \tilde{p}_{zs}$
Vector potential	$q_s \boldsymbol{A} = m_s v_{\text{th},s} \tilde{\boldsymbol{A}}_s$
Scalar Potential	$\tilde{\phi}_s = q_s \beta_s \phi$
Bulk rectilinear flows	$v_{\text{th},s} \tilde{U}_{zs} = U_{zs}, \quad v_{\text{th},s} \tilde{V}_{zs} = V_{zs}$
Bulk angular frequency	$\tau v_{\text{th},s} \tilde{\omega}_s = \omega_s$
Particle position (radial)	$\tau r = \tilde{r}$
Particle velocity	$\boldsymbol{v} = v_{\text{th},s} \tilde{\boldsymbol{v}}_s$

5.4.1.1 Zeroth Order Moments

The number density of species s is given by the zeroth moment of the DF;

$$n_s = \int f_s d^3 v_s = \frac{n_{0s}}{(\sqrt{2\pi})^3} \int e^{-\tilde{H}_s} \left(e^{\tilde{U}_{zs} \tilde{p}_{zs}} e^{\tilde{\omega}_s \tilde{p}_{\theta s}} + C_s e^{\tilde{V}_{zs} \tilde{p}_{zs}} \right) d^3 \tilde{v}_s \tag{5.21}$$

$$= \frac{n_{0s}}{(\sqrt{2\pi})^2} e^{-\tilde{\phi}_s} \Bigg[e^{(\tilde{U}_{zs}^2 + \tilde{r}^2 \tilde{\omega}_s^2)/2} e^{\tilde{U}_{zs} \tilde{A}_{zs}} e^{\tilde{\omega}_s \tilde{r} \tilde{A}_{\theta s}} \int_{-\infty}^{\infty} e^{-(\tilde{v}_{zs} - \tilde{U}_{zs})^2/2} d\tilde{v}_{zs} \times$$

$$\int_{-\infty}^{\infty} e^{-(\tilde{v}_{\theta s} - \tilde{\omega}_s \tilde{r})^2/2} d\tilde{v}_{\theta s} + C_s \sqrt{2\pi} e^{\tilde{V}_{zs}^2/2} e^{\tilde{V}_{zs} \tilde{A}_{zs}} \int_{-\infty}^{\infty} e^{-(\tilde{v}_{zs} - \tilde{V}_{zs})^2/2} d\tilde{v}_{zs} \Bigg]$$

$$= n_{0s} e^{-\tilde{\phi}_s} \left[e^{(\tilde{U}_{zs}^2 + \tilde{r}^2 \tilde{\omega}_s^2)/2} e^{\tilde{U}_{zs} \tilde{A}_{zs}} e^{\tilde{\omega}_s \tilde{r} \tilde{A}_{\theta s}} + C_s e^{\tilde{V}_{zs}^2/2} e^{\tilde{V}_{zs} \tilde{A}_{zs}} \right] \tag{5.22}$$

We take the following sum to calculate the charge density,

$$\sigma = \sum_s q_s n_s = \sum_s n_{0s} q_s e^{-\tilde{\phi}_s} \left[e^{(\tilde{U}_{zs}^2 + \tilde{r}^2 \tilde{\omega}_s^2)/2} e^{\tilde{U}_{zs} \tilde{A}_{zs}} e^{\tilde{\omega}_s \tilde{r} \tilde{A}_{\theta s}} + C_s e^{\tilde{V}_{zs}^2/2} e^{\tilde{V}_{zs} \tilde{A}_{zs}} \right] \tag{5.23}$$

5.4.1.2 First Order Moments

We take the v_z moment of the DF to calculate the z—component of the bulk velocity,

$$V_{zs} = \frac{v_{\text{th},s}^4}{n_s} \int \tilde{v}_{zs} f_s d^3\tilde{v}_s,$$
$$= \frac{v_{\text{th},s}}{n_s} \frac{n_{0s}}{(\sqrt{2\pi})^2} e^{-\tilde{\phi}_s} \Big[e^{(\tilde{U}_{zs}^2 + \tilde{r}^2\tilde{\omega}_s^2)/2} e^{\tilde{U}_{zs}\tilde{A}_{zs}} e^{\tilde{\omega}_s \tilde{r} \tilde{A}_{\theta s}} \int_{-\infty}^{\infty} \tilde{v}_{zs} e^{-(\tilde{v}_{zs} - \tilde{U}_{zs})^2/2} d\tilde{v}_{zs} \times$$
$$\int_{-\infty}^{\infty} e^{-(\tilde{v}_{\theta s} - \tilde{\omega}_s \tilde{r})^2/2} d\tilde{v}_{\theta s} + C_s \sqrt{2\pi} e^{\tilde{V}_{zs}^2/2} e^{\tilde{V}_{zs}\tilde{A}_{zs}} \int_{-\infty}^{\infty} \tilde{v}_{zs} e^{-(\tilde{v}_{zs} - \tilde{V}_{zs})^2/2} d\tilde{v}_{zs} \Big]$$
$$= \frac{n_{0s} v_{\text{th},s}}{n_s} e^{-\tilde{\phi}_s} \Big[\tilde{U}_{zs} e^{\tilde{U}_{zs}\tilde{A}_{zs}} e^{(\tilde{U}_{zs}^2 + \tilde{r}^2\tilde{\omega}_s^2)/2} e^{\tilde{\omega}_s \tilde{r} \tilde{A}_{\theta s}} + \tilde{V}_{zs} C_s e^{\tilde{V}_{zs}^2/2} e^{\tilde{V}_{zs}\tilde{A}_{zs}} \Big], \tag{5.24}$$

for n_s the number density. We take the following sum to calculate the z—component of the current density,

$$j_z = \sum_s q_s n_s V_{zs} = \sum_s n_{0s} q_s v_{\text{th},s} e^{-\tilde{\phi}_s} \times$$
$$\left(\tilde{U}_{zs} e^{\tilde{U}_{zs}\tilde{A}_{zs}} e^{(\tilde{U}_{zs}^2 + \tilde{r}^2\tilde{\omega}_s^2)/2} e^{\tilde{\omega}_s \tilde{r} \tilde{A}_{\theta s}} + \tilde{V}_{zs} C_s e^{\tilde{V}_{zs}^2/2} e^{\tilde{V}_{zs}\tilde{A}_{zs}} \right). \tag{5.25}$$

By taking the v_θ moment of the DF we can calculate the θ—component of the bulk velocity,

$$V_{\theta s} = \frac{v_{\text{th},s}^4}{n_s} \int \tilde{v}_{\theta s} f_s d^3\tilde{v}_s = \frac{v_{\text{th},s}}{n_s} \frac{n_{0s}}{(\sqrt{2\pi})^2} e^{-\tilde{\phi}_s} \Big[e^{(\tilde{U}_{zs}^2 + \tilde{r}^2\tilde{\omega}_s^2)/2} e^{\tilde{U}_{zs}\tilde{A}_{zs}} e^{\tilde{\omega}_s \tilde{r} \tilde{A}_{\theta s}} \times$$
$$\int_{-\infty}^{\infty} e^{-(\tilde{v}_{zs} - \tilde{U}_{zs})^2/2} d\tilde{v}_{zs} \int_{-\infty}^{\infty} \tilde{v}_{\theta s} e^{-(\tilde{v}_{\theta s} - \tilde{\omega}_s \tilde{r})^2/2} d\tilde{v}_{\theta s}$$
$$= \frac{\tilde{r}\tilde{\omega}_s n_{0s} v_{\text{th},s} e^{-\tilde{\phi}_s}}{n_s} e^{(\tilde{U}_{zs}^2 + \tilde{r}^2\tilde{\omega}_s^2)/2} e^{\tilde{U}_{zs}\tilde{A}_{zs}} e^{\tilde{\omega}_s \tilde{r} \tilde{A}_{\theta s}}, \tag{5.26}$$

for n_s the number density. This gives the θ—component of the current density,

$$j_\theta = \sum_s q_s n_s V_{\theta s} = \sum_s n_{0s} q_s v_{\text{th},s} \tilde{r}\tilde{\omega}_s e^{-\tilde{\phi}_s} e^{\tilde{U}_{zs}\tilde{A}_{zs}} e^{(\tilde{U}_{zs}^2 + \tilde{r}^2\tilde{\omega}_s^2)/2} e^{\tilde{\omega}_s \tilde{r} \tilde{A}_{\theta s}}. \tag{5.27}$$

5.4.2 *Maxwell's Equations: Fixing the Parameters of the DF*

By insisting on a specific magnetic field configuration (the GH+B field) we have made a statement on the macroscopic physics. In searching for the equilibrium DF, we are trying to understand the microscopic physics. In this sense we are tackling an 'inverse problem'. Once an assumption on the form of the DF is made then – should the assumed form be able to reproduce the correct moments – this inverse problem reduces to establishing the relationships between the microscopic and macroscopic parameters of the equilibrium. In this Section we 'fix' the free parameters of the DF in Eq. (5.20), such that Maxwell's equations are satisfied;

$$\nabla \cdot \boldsymbol{E} = \frac{1}{\varepsilon_0} \sum_s q_s \int f_s d^3 v, \tag{5.28}$$

$$\nabla \times \boldsymbol{B} = \mu_0 \sum_s q_s \int \boldsymbol{v} f_s d^3 v. \tag{5.29}$$

Note that the solenoidal constraint and Faraday's law are automatically satisfied for the GH+B field in equilibrium, since $\boldsymbol{B} = \nabla \times \boldsymbol{A}$ implies that $\nabla \cdot \boldsymbol{B} = 0$ and $\boldsymbol{E} = -\nabla \phi$ implies that $\nabla \times \boldsymbol{E} = \boldsymbol{0} = -\frac{\partial \boldsymbol{B}}{\partial t}$.

5.4.2.1 Ampère's Law

In Sect. 5.4.1.2 we have calculated the j_z current density, found by summing first order moments in v_z of the DF. We now substitute in the macroscopic expressions for $j_z(\tilde{r})$, $A_\theta(\tilde{r})$ and $A_z(\tilde{r})$ from (5.19) and (5.17) into the expression for the j_z current density of Eq. (5.25). After this substitution, we can calculate a $\phi(r)$ that makes the system consistent. The substitution of the known expressions for j_z, A_z and A_θ gives

$$\begin{aligned} j_z(\tilde{r}) = \frac{2\tau B_0}{\mu_0} \frac{1}{(1+\tilde{r}^2)^2} &= \sum_s n_{0s} q_s v_{\mathrm{th},s} e^{-q_s \beta_s \phi} \times \\ &\left(\tilde{U}_{zs} e^{(\tilde{U}_{zs}^2 + \tilde{r}^2 \tilde{\omega}_s^2)/2 - \mathrm{sgn}(q_s)\tilde{\omega}_s \tilde{r}^2 k/\delta_s} \left(1+\tilde{r}^2\right)^{\mathrm{sgn}(q_s)(\tilde{\omega}_s - \tilde{U}_{zs})/(2\delta_s)} \right. \\ &\left. + \tilde{V}_{zs} C_s e^{\tilde{V}_{zs}^2/2} \left(1+\tilde{r}^2\right)^{-\mathrm{sgn}(q_s)\tilde{V}_{zs}/(2\delta_s)} \right) \\ &= \text{"ion terms"} + \text{"electron terms"} \end{aligned} \tag{5.30}$$

In order to satisfy the above equality we can construct a solution by introducing a 'separation constant' $\gamma_1 \neq 0, 1$. We multiply the above equation by $(1+\tilde{r}^2)^2$ which makes the left-hand side constant, whilst the right-hand side is a sum of two (sets of) terms, one depending on ion parameters and the second depending on electron parameters. Then we can define γ_1 by

$$\frac{2\tau B_0}{\mu_0} = \underbrace{\frac{2\tau B_0}{\mu_0}(1-\gamma_1)}_{\text{ion terms}} + \underbrace{\frac{2\tau B_0}{\mu_0}\gamma_1}_{\text{electron terms}}, \tag{5.31}$$

associating the 'ion term' with the first term on the right-hand side of (5.31), and the 'electron term' with the second term on the right-hand side of (5.31). After some algebra we can rearrange these two associations to give two expressions for the scalar potential, one in terms of the ion parameters, and one in terms of the electron parameters:

$$\phi(r) = \frac{1}{q_i\beta_i}\ln\left\{\frac{\mu_0 n_{0i} q_i v_{\text{th},i}}{2\tau B_0(1-\gamma_1)}\times\right.$$
$$\left[\tilde{U}_{zi}e^{(\tilde{U}_{zi}^2+\tilde{r}^2\tilde{\omega}_i^2)/2-\tilde{\omega}_i\tilde{r}^2k/\delta_i}\left(1+\tilde{r}^2\right)^{2+(\tilde{\omega}_i-\tilde{U}_{zi})/(2\delta_i)}\right.$$
$$\left.\left.+\tilde{V}_{zi}C_i e^{\tilde{V}_{zi}^2/2}\left(1+\tilde{r}^2\right)^{2-\tilde{V}_{zi}/(2\delta_i)}\right]\right\}$$
$$\phi(r) = \frac{1}{q_e\beta_e}\ln\left\{\frac{\mu_0 n_{0e} q_e v_{\text{th},e}}{2\tau B_0\gamma_1}\left[\tilde{U}_{ze}e^{(\tilde{U}_{ze}^2+\tilde{r}^2\tilde{\omega}_e^2)/2+\tilde{\omega}_e\tilde{r}^2k/\delta_e}\left(1+\tilde{r}^2\right)^{2-(\tilde{\omega}_e-\tilde{U}_{ze})/(2\delta_e)}\right.\right.$$
$$\left.\left.+\tilde{V}_{ze}C_e e^{\tilde{V}_{ze}^2/2}\left(1+\tilde{r}^2\right)^{2+\tilde{V}_{ze}/(2\delta_e)}\right]\right\}$$

The two values of the scalar potential above must be made identical by a suitable choice of relationships between the ion and electron parameters. Given enough freedom in parameter space, we could say that the z component of Ampère's Law is *implicitly* solved by the above equations, in that one just needs to choose a consistent set of parameters. However, we seek a solution in an *explicit* sense.

In order to make progress we non-dimensionalise the above equations by multiplying both sides by $e\beta_r$ with

$$\beta_r = \frac{\beta_i\beta_e}{\beta_e+\beta_i}.$$

Once this is done we can write the scalar potential in the form

$$e\beta_r\phi(r) = \ln\left\{[\text{ion terms}]^{\frac{e\beta_r}{q_i\beta_i}}\right\}, \tag{5.32}$$
$$e\beta_r\phi(r) = \ln\left\{[\text{electron terms}]^{\frac{e\beta_r}{q_e\beta_e}}\right\}. \tag{5.33}$$

Specifically, Eqs. (5.32) and (5.33) require the equality of the arguments of the logarithm to hold in order for a meaningful solution to be obtained for the scalar potential. A first step towards this is made by requiring consistent powers of the $1+\tilde{r}^2$ 'profile' in the right-hand side of the above expression to allow factorisation. Hence

$$(\tilde{\omega}_i - \tilde{U}_{zi})/(2\delta_i) = -\tilde{V}_{zi}/(2\delta_i), \quad -(\tilde{\omega}_e - \tilde{U}_{ze})/(2\delta_e) = \tilde{V}_{ze}/(2\delta_e),$$
$$\Longrightarrow \tilde{\omega}_i = \tilde{U}_{zi} - \tilde{V}_{zi}, \quad \tilde{\omega}_e = \tilde{U}_{ze} - \tilde{V}_{ze}, \tag{5.34}$$

and hence the rigid-rotation, $\tilde{\omega}_s$, is fixed by the difference of the rectilinear motion, $\tilde{U}_{zs} - \tilde{V}_{zs}$. On top of this, we require that the power of the $1+\tilde{r}^2$ 'profile' on the right-hand side is the same for both the ions and electrons, thus

$$\frac{e\beta_r}{q_i\beta_i}\left(2 - \tilde{V}_{zi}/(2\delta_i)\right) = \mathcal{E} = \frac{e\beta_r}{q_e\beta_e}\left(2 + \tilde{V}_{ze}/(2\delta_e)\right). \tag{5.35}$$

This condition seems to be a statement on an average potential energy associated with the particles. Once more to allow factorisation of the $1 + \tilde{r}^2$ 'profile', we insist that net $\exp(r^2)$ terms cancel, i.e.

$$\frac{\tilde{\omega}_i}{2} = \frac{k}{\delta_i} > 0, \quad \frac{\tilde{\omega}_e}{2} = -\frac{k}{\delta_e} < 0. \tag{5.36}$$

The physical meaning of this condition seems to be that the frequencies of the rigid rotor for each species are matched according to the relevant magnetisation, and the background field magnitude. The remaining task is to ensure equality of the 'coefficients'

$$\left\{ \frac{1}{4\delta_i(1-\gamma_1)} \frac{n_{0i} m_i v_{\text{th},i}^2}{B_0^2/(2\mu_0)} \left[\tilde{U}_{zi} e^{\tilde{U}_{zi}^2/2} + \tilde{V}_{zi} C_i e^{\tilde{V}_{zi}^2/2} \right] \right\}^{\frac{e\beta_r}{q_i \beta_i}} = \mathcal{D}$$

$$= \left\{ -\frac{1}{4\delta_e \gamma_1} \frac{n_{0e} m_e v_{\text{th},e}^2}{B_0^2/(2\mu_0)} \left[\tilde{U}_{ze} e^{\tilde{U}_{ze}^2/2} + \tilde{V}_{ze} C_e e^{\tilde{V}_{ze}^2/2} \right] \right\}^{\frac{e\beta_r}{q_e \beta_e}} \tag{5.37}$$

These seem to be conditions on the ratios of the energy densities associated with the bulk rectilinear motion and the magnetic field respectively. Thus far we have 8 constraints and 12 unknowns ($\tilde{U}_{zs}, \tilde{V}_{zs}, \tilde{\omega}_s, C_s, n_{0s}, \beta_s$), given fixed characteristic macroscopic parameters of the equilibrium; B_0, τ, and k. We can now write down an expression for ϕ that explicitly solves the z component of Ampère's law;

$$\phi(\tilde{r}) = \frac{1}{e\beta_r} \mathcal{E} \ln\left(1 + \tilde{r}^2\right) + \phi(0), \tag{5.38}$$

with

$$\phi(0) = \frac{1}{e\beta_r} \ln \mathcal{D}.$$

Clearly, we require that $\mathcal{D} > 0$ for the expression above to make sense. It is clear that the sign of γ_1 could, in principle, affect the sign of $\mathcal{D}$. It is seen from (5.37) that positivity of $\mathcal{D}$ implies that

$$\frac{1}{1-\gamma_1} \left[\tilde{U}_{zi} e^{\tilde{U}_{zi}^2/2} + \tilde{V}_{zi} C_i e^{\tilde{V}_{zi}^2/2} \right] > 0, \tag{5.39}$$

$$\frac{1}{\gamma_1} \left[\tilde{U}_{ze} e^{\tilde{U}_{ze}^2/2} + \tilde{V}_{ze} C_e e^{\tilde{V}_{ze}^2/2} \right] < 0. \tag{5.40}$$

By rearranging the above inequalities to make C_s the subject, it can be seen after some algebra that positivity of $\mathcal{D}$ and C_s is guaranteed when

$$\gamma_1 > 1, \quad \text{sgn}(\tilde{U}_{zs}) = -\text{sgn}(\tilde{V}_{zs}).$$

Note that these conditions are sufficient, but not necessary, i.e. it is possible to have $\mathcal{D} > 0$ and $C_s > 0$ for any value of $\gamma_1 \neq 0, 1$, and even for $\text{sgn}(\tilde{U}_{zs}) = \text{sgn}(\tilde{V}_{zs})$ in the case of $\gamma_1 < 0$.

Thus far we have only considered the j_z component, and it is premature to consider all components of Ampère's Law satisfied. Let us move on to consider the θ component. In a process similar to that above, we substitute in the macroscopic expressions for $j_\theta(\tilde{r})$, $A_\theta(\tilde{r})$ and $A_z(\tilde{r})$ for the GH+B field into the expression for the j_θ current density of Eq. (5.27) in Sect. 5.4.1.2. After this substitution, we can once more calculate the ϕ that makes the system consistent. The substitution gives

$$j_\theta = \frac{2\tau B_0}{\mu_0} = \sum_s n_{0s} q_s v_{\text{th},s} \tilde{\omega}_s e^{-q_s \beta_s \phi} \times$$
$$e^{(\tilde{U}_{zs}^2 + \tilde{r}^2 \tilde{\omega}_s^2)/2 - \text{sgn}(q_s)\tilde{\omega}_s \tilde{r}^2 k/\delta_s} \left(1 + \tilde{r}^2\right)^{2 + \text{sgn}(q_s)(\tilde{\omega}_s - \tilde{U}_{zs})/(2\delta_s)} \tag{5.41}$$

Using the parameter relations as above, we determine that the scalar potential is again given in the form of (5.38),

$$\phi(\tilde{r}) = \frac{1}{e\beta_r} \mathcal{E} \ln\left(1 + \tilde{r}^2\right) + \phi(0).$$

Hence, this form of the scalar potential is consistent provided

$$\left[\frac{1}{1-\gamma_2} \frac{1}{4\delta_i} \frac{n_{0i} m_i v_{\text{th},i} \omega_i/\tau}{B_0^2/(2\mu_0)} e^{\tilde{U}_{zi}^2/2}\right]^{\frac{e\beta_r}{q_i\beta_i}} = \mathcal{D} = \left[-\frac{1}{\gamma_2} \frac{1}{4\delta_e} \frac{n_{0e} m_e v_{\text{th},e} \omega_e/\tau}{B_0^2/(2\mu_0)} e^{\tilde{U}_{ze}^2/2}\right]^{\frac{e\beta_r}{q_e\beta_e}} \tag{5.42}$$

for $\gamma_2 \neq 1$ another separation constant. These seem to be conditions on the ratios of the energy densities associated with the bulk rotation and the magnetic field respectively. This has added two more constraints.

Once again we must ensure that $\mathcal{D} > 0$. Since $\omega_e < 0$, the right-hand side of the above equation implies that $\gamma_2 > 0$ to ensure that $\mathcal{D} > 0$. Whilst the left-hand side implies that $\gamma_2 < 1$ for positivity of $\mathcal{D}$ since $\omega_i > 0$. Hence we can say that for positivity

$$0 < \gamma_2 < 1.$$

We can now consider Ampère's Law satisfied, given a ϕ that solves Poisson's equation. That is to say that we have satisfied the equation

$$\left(\sum_s q_s \int \boldsymbol{v} f_s d^3 v = \right) \boldsymbol{j}_{\text{micro}}(\phi, \boldsymbol{A}) = \boldsymbol{j}_{\text{macro}}(r) \left(= \frac{1}{\mu_0} \nabla \times \boldsymbol{B}\right),$$
$$\text{s.t.} \quad \phi(\tilde{r}) = \frac{1}{e\beta_r} \mathcal{E} \ln\left(1 + \tilde{r}^2\right) + \phi(0) \quad \text{and} \quad \boldsymbol{A} = \boldsymbol{A}_{GH+B},$$

with $\boldsymbol{A}_{GH+B}$ defined by Eq. (5.17). As a result, the problem of consistency is now shifted to solving Poisson's Equation, where the remaining degrees of freedom lie.

5.4.2.2 Poisson's Equation

The final step in 'self-consistency' is to solve Poisson's Equation. Frequently in such equilibrium studies, this step is replaced by satisfying quasineutrality and in essence solving a first order approximation of Poisson's equation, see for example Schindler (2007), Harrison and Neukirch (2009b), Tasso and Throumoulopoulos (2014) and Sect. 1.1.3 of this thesis. Here we solve Poisson's equation exactly, i.e. to all orders. Poisson's equation in cylindrical coordinates with only radial dependence gives

$$\nabla \cdot \boldsymbol{E} = -\frac{1}{r}\frac{\partial}{\partial r}\left(r\frac{\partial \phi}{\partial r}\right) = \frac{\sigma}{\varepsilon_0}. \tag{5.43}$$

The electric field is calculated as $\boldsymbol{E} = -\nabla\phi$, giving

$$E_r = -\partial_r \phi = -\frac{2\tau\mathcal{E}}{e\beta_r}\frac{\tilde{r}}{(1+\tilde{r}^2)}. \tag{5.44}$$

We can now take the divergence of the electric field $\nabla \cdot \boldsymbol{E} = \tau\tilde{r}^{-1}\partial_{\tilde{r}}(\tilde{r}E_r)$ and so

$$\nabla \cdot \boldsymbol{E} = -\frac{4\tau^2\mathcal{E}}{e\beta_r}\frac{1}{(1+\tilde{r}^2)^2} \implies \sigma = -\frac{4\varepsilon_0\tau^2\mathcal{E}}{e\beta_r}\frac{1}{(1+\tilde{r}^2)^2}. \tag{5.45}$$

This gives a non-zero net charge—per unit length in z—of

$$\mathcal{Q} = \int_{\theta=0}^{\theta=2\pi}\int_{r=0}^{r=\infty} \sigma\, r\, dr\, d\theta = -\frac{4\pi\varepsilon_0\mathcal{E}}{e\beta_r}. \tag{5.46}$$

The charge density derived in Eq. (5.45) must equal the charge density calculated by taking the zeroth moment of the DF. The expression for the charge density calculated in (5.23) gives

$$\begin{aligned}
\sigma &= \sum_s n_{0s}q_s e^{-q_s\beta_s\phi} \times \\
&\quad \left(e^{(\tilde{U}_{zs}^2+\tilde{r}^2\tilde{\omega}_s^2)/2}e^{\tilde{U}_{zs}\tilde{A}_{zs}}e^{\tilde{\omega}_s\tilde{r}\tilde{A}_{\theta s}} + C_s e^{(\tilde{U}_{zs}-\tilde{\omega}_s)^2/2}e^{(\tilde{U}_{zs}-\tilde{\omega}_s)\tilde{A}_{zs}}\right), \\
&= \sum_s n_{0s}q_s e^{-q_s\beta_s\phi} \times \\
&\quad \left(1+\tilde{r}^2\right)^{\mathrm{sgn}(q_s)(\tilde{\omega}_s-\tilde{U}_{zs})/(2\delta_s)}\left(e^{\tilde{U}_{zs}^2/2} + C_s e^{(\tilde{U}_{zs}-\tilde{\omega}_s)^2/2}\right), \\
&= \frac{1}{\left(1+\tilde{r}^2\right)^2}\sum_s n_{0s}q_s \mathcal{D}^{-\frac{q_s\beta_s}{e\beta_r}}\left(e^{\tilde{U}_{zs}^2/2} + C_s e^{(\tilde{U}_{zs}-\tilde{\omega}_s)^2/2}\right).
\end{aligned} \tag{5.47}$$

The second equality is found by substituting the form of the vector potential from Eq. (5.17), and the final equality is reached by using the conditions derived in Eqs. (5.34)–(5.38).

We can now match Eqs. (5.45) and (5.47) to get

$$(\sigma(0) =) - \frac{4\varepsilon_0\tau^2\mathcal{E}}{e\beta_r} = \sum_s n_{0s}q_s\mathcal{D}^{-\frac{q_s\beta_s}{e\beta_r}}\left(e^{\tilde{U}_{zs}^2/2} + C_s e^{\tilde{V}_{zs}^2/2}\right). \tag{5.48}$$

We now have 12 physical parameters ($\tilde{U}_{zs}$, $\tilde{V}_{zs}$, $\tilde{\omega}_s$, C_s, n_{0s}, β_s) with 11 constraints (5.34–5.37), (5.42) and (5.48). For example, if one picks B_0, τ, k and one microscopic parameter, say β_i, then the remaining parameters of the equilibrium, ($\tilde{U}_{zs}$, $\tilde{V}_{zs}$, $\tilde{\omega}_s$, C_s, n_{0s}, β_e), are now determined. One could of course choose the values of a different set of parameters, and determine those that remain by using the constraints derived. Note that whilst the constants $\gamma_1 \neq 0, 1$ and $0 < \gamma_2 < 1$ are system parameters, they are not physically meaningful as they only represent a change in the gauge of the scalar potential.

5.5 Analysis of the Equilibrium

5.5.1 *Non-neutrality and the Electric Field*

It is seen from Eqs. (5.45) and (5.46) that basic electrostatic properties of the equilibrium described by f_s are encoded in $\mathcal{E}$. The equilibrium is electrically neutral only when $\mathcal{E} = 0$, and non-neutral otherwise. Specifically, there is net negative charge when $\mathcal{E} > 0$, and net positive charge when $\mathcal{E} < 0$. This net charge is finite in the (r, θ) plane and given by $\mathcal{Q}$ in Eq. (5.46).

Physically, the sign of $\mathcal{E}$ seems to be related to the respective magnitudes of the bulk rotation frequencies, $\tilde{\omega}_s$. From Eqs. (5.34) and (5.35) we see that $\mathcal{E} > 0$ implies that

$$\tilde{\omega}_i > \omega_i^\star = \tilde{U}_{zi} - 4\delta_i,$$
$$|\tilde{\omega}_e| < \omega_e^\star = -\tilde{U}_{ze} - 4\delta_e,$$

and $\mathcal{E} < 0$ implies that

$$\tilde{\omega}_i < \omega_i^\star = \tilde{U}_{zi} - 4\delta_i,$$
$$|\tilde{\omega}_e| > \omega_e^\star = -\tilde{U}_{ze} - 4\delta_e.$$

Hence, $\mathcal{E} > 0$ is seen to occur for 'sufficiently large' bulk ion rotation frequencies, and 'sufficiently small' (in magnitude) bulk electron rotation frequencies. A positive $\mathcal{E}$ corresponds to an electric field directed radially 'inwards'. This seems to make sense physically, by the following argument. A 'larger' ($\tilde{\omega}_i > \omega_i^\star$) bulk ion rotation frequency gives a 'larger' centrifugal force (in the co-moving frame), and a 'smaller' ($|\tilde{\omega}_e| < \omega_e^\star$) bulk electron rotation frequency gives a 'smaller' centrifugal force (in the co-moving frame). For a dynamic interpretation, at a fixed r, the ions are forced to a slightly larger radius than the electrons, i.e. a charge separation manifests on small scales. This charge separation results in an inward electric field, $E_r < 0$. An equally valid interpretation is to say that for an equilibrium to exist, an electric field must exist to counteract the differences in the forces associated with the bulk ion and electron rotational flows. This effect is represented in Fig. 5.1a.

In a similar manner, $\mathcal{E} < 0$ is seen to occur for 'sufficiently small' ($\tilde{\omega}_i < \omega_i^\star$) bulk ion rotation frequencies, and 'sufficiently large' ($|\tilde{\omega}_e| > \omega_e^\star$) bulk electron rotation frequencies. A negative $\mathcal{E}$ corresponds to an electric field directed radially 'outwards'. We can then interpret these result physically, in a manner like that above. This effect is represented in Fig. 5.1b.

Finally, we can interpret the neutral case, $\mathcal{E} = 0$, as the intermediary between the two circumstances considered above. That is to say that the equilibrium is neutral when the bulk rotation flows are just matched accordingly, such that there is no charge separation and hence no electric field.

5.5.2 *The Equation of State and the Plasma Beta*

For certain considerations, e.g. the solar corona, it would be advantageous if the DF had the capacity to describe plasmas with sub-unity values of the plasma beta: the ratio of the thermal energy density to the magnetic energy density

$$\beta_{pl}(\tilde{r}) = \frac{2\mu_0 k_B}{B^2} \sum_s n_s T_s. \tag{5.49}$$

For our configuration, the number density is seen to be proportional to the rr component of the pressure tensor, $P_{rr,s} = n_s k_B T_s$. This is demonstrated by the following calculation. In order to calculate P_{rr}, we must consider the integral

$$P_{rr} = \sum_s m_s \int_{-\infty}^{\infty} w_{rs}\, w_{rs}\, f_s\, d^3v. \tag{5.50}$$

However, we do not have to consider a bulk velocity in the r direction here ($V_{rs} = 0$), since f_s is an even function of v_r. Using the fact that

$$\int_{-\infty}^{\infty} v_r^2 e^{-v_r^2/(2v_{\text{th},s}^2)} dv_r = v_{\text{th},s}^2 \int_{-\infty}^{\infty} e^{-v_r^2/(2v_{\text{th},s}^2)} dv_r,$$

and by consideration of Eq. (5.50) and the number density, we see that

$$P_{rr,s} = m_s v_{\mathrm{th},s}^2 n_s, \tag{5.51}$$

that is to say that $k_B T_s = m_s v_{\mathrm{th},s}^2$. Note that if $n_i = n_e := n$ and hence $\mathcal{E} = 0$ (neutrality), then we have an equation of state given by

$$P_{rr} = \frac{\beta_e + \beta_i}{\beta_e \beta_i} n.$$

This resembles expressions found in the Cartesian case, in Channell (1976), Neukirch et al. (2009), Allanson et al. (2015) for example. Incidentally, we can use the connection between n_s and P_{rr} to give an expression for the β_{pl} that is perhaps more typically seen,

$$\beta_{pl}(\tilde{r}) = \frac{2\mu_0}{B^2} \sum_s P_{rr,s}.$$

The square magnitude of the magnetic field (Eq. 5.18) is given by

$$B^2 = \frac{B_0^2}{(1+\tilde{r}^2)} \left(1 - 4k + 4k^2(1+\tilde{r}^2)\right).$$

Using the number density from Eq. (5.22) in the definition of the plasma beta from Eq. (5.49), as well as the equilibrium conditions (5.34)–(5.38) gives

$$\beta_{pl}(\tilde{r}) = \frac{2\mu_0}{B_0^2(1+\tilde{r}^2)\left(1 - 4k + 4k^2(1+\tilde{r}^2)\right)} \times \sum_s \frac{n_{0s}}{\beta_s} \mathcal{D}^{-\frac{q_s \beta_s}{e\beta_r}} \left(e^{\tilde{U}_{zs}^2/2} + C_s e^{\tilde{V}_{zs}^2/2}\right). \tag{5.52}$$

It is not immediately obvious from the above equation what values β_{pl} can have. However it is readily seen that as $\tilde{r} \to \infty$ then $\beta_{pl} \to 0$, essentially since the number density is vanishing at large radii. On the central axis of the tube we see that

$$\beta_{pl}(0) = \frac{2\mu_0}{B_0^2\left(1 - 4k + 4k^2\right)} \times \sum_s \frac{n_{0s}}{\beta_s} \mathcal{D}^{-\frac{q_s \beta_s}{e\beta_r}} \left(e^{\tilde{U}_{zs}^2/2} + C_s e^{\tilde{V}_{zs}^2/2}\right), \tag{5.53}$$

suggesting that for a suitable choice of parameters, it should be possible to attain any value of β_{pl} on the axis.

5.5.3 Origin of Terms in the Equation of Motion

It could be instructive to now consider the individual terms in the equation of motion for this equilibrium, Eq. (5.11), and repeated here,

$$(\nabla \cdot \boldsymbol{P})_r = (\boldsymbol{j} \times \boldsymbol{B})_r + \sigma E_r - \mathcal{F}_\mathrm{c} \cdot \hat{\boldsymbol{e}}_r.$$

We will seek to see if, at least mathematically, that certain terms have their origin in other particular terms in the equation, and what these are. Rather than this suggesting 'what balances what', it is an attempt to see the physical origin of the forces, i.e. *which forces arise from which system configurations?*

5.5.3.1 Centripetal Forces and Non-inertial Motion

Let's first consider the divergence of the pressure, Eq. (5.9), and repeated here

$$(\nabla \cdot \boldsymbol{P})_r = \frac{1}{r}\frac{\partial}{\partial r}(r P_{rr}) - \frac{P_{\theta\theta}}{r}.$$

As mentioned in Sect. 5.3.3, $P_{\theta\theta} = \pi_{\theta\theta} - \sum_s n_s V_{\theta s}^2$, since

$$\begin{aligned} P_{\theta\theta} &= \sum_s \int (V_{\theta s} - v_\theta)^2 f_s d^3 v, \\ &= \sum_s \left[n_s V_{\theta s}^2 - 2 n_s V_{\theta s}^2 + \int v_\theta^2 f_s d^3 v \right], \\ &= \sum_s \left[\int v_\theta^2 f_s d^3 v - n_s V_{\theta s}^2 \right], \\ &= \pi_{\theta\theta} - \sum_s n_s V_{\theta s}^2 = \pi_{\theta\theta} + \mathcal{F}_c \cdot r\hat{\boldsymbol{e}}_r. \end{aligned}$$

Hence the centripetal forces, $\mathcal{F}_c = -\frac{1}{r}\sum_s \rho_s V_{\theta s}^2 \hat{\boldsymbol{e}}_r$ are seen to have their origin in the terms in $P_{\theta\theta}/r$, from $\nabla \cdot \boldsymbol{P}$. This seems to say that in a lab frame, the centripetal forces arise from the stresses associated with the differences between the particle and bulk velocities, i.e. the $\int v_\theta V_{\theta s} f_s d^3 v$ terms. So far we have accounted for the following terms,

$$\underbrace{\frac{1}{r}\frac{\partial}{\partial r}(r P_{rr}) - \frac{1}{r}\pi_{\theta\theta} + \frac{1}{r}\sum_s n_s V_{\theta s}^2}_{\text{"Derivatives" of potentials}} = \underbrace{(\boldsymbol{j} \times \boldsymbol{B})_r + \sigma E_r - \mathcal{F}_\mathrm{c} \cdot \hat{\boldsymbol{e}}_r}_{\text{Forces}}.$$

5.5.3.2 Electric Fields and Pressure Gradients

We now consider the P_{rr} terms. Using the 'equation of state' (5.51), and the n_s implicit from Eq. (5.47) we see that

$$\frac{1}{r}\frac{\partial}{\partial r}(r P_{rr}) = \frac{1}{r}\frac{\partial}{\partial r}\left[r\left(\frac{n_i}{\beta_i}+\frac{n_e}{\beta_e}\right)\right] \propto \frac{1}{\tilde{r}}\frac{\partial}{\partial \tilde{r}}\left[\tilde{r}\frac{1}{(1+\tilde{r}^2)^2}\right],$$
$$= \underbrace{\frac{1}{\tilde{r}}\frac{1}{(1+\tilde{r}^2)^2}}_{\propto P_{rr}/r} - \underbrace{\frac{4\tilde{r}}{(1+\tilde{r}^2)^3}}_{\propto \partial P_{rr}/\partial r} \quad (5.54)$$

We can see from Eqs. (5.44), (5.47) and (5.48), that

$$\sigma E_r = \frac{8\epsilon_0 \tau^3 \mathcal{E}^2}{e^2 \beta_r^2}\frac{\tilde{r}}{(1+\tilde{r}^2)^3}.$$

Hence the electric fields have their origin in the density/pressure gradients $\partial P_{rr}/\partial r$, and we have accounted for the following terms,

$$\underbrace{\frac{1}{r}P_{rr} + \frac{\partial}{\partial r}\sum_s \frac{n_s}{\beta_s} - \frac{1}{r}\pi_{\theta\theta} + \frac{1}{r}\sum_s n_s V_{\theta s}^2}_{\text{"Derivatives" of potentials}} = \underbrace{(\boldsymbol{j}\times\boldsymbol{B})_r + \sigma E_r - \boldsymbol{\mathcal{F}}_{\mathrm{c}}\cdot\hat{\boldsymbol{e}}_r}_{\text{Forces}}.$$

5.5.3.3 'Lorentz Forces' and $\pi_{\theta\theta}$

Using the definition of the DF (Eq. 5.20), let's now consider the form of $\pi_{\theta\theta}/r$,

$$-\frac{1}{r}\pi_{\theta\theta} = -\frac{1}{r}\sum_s \int v_\theta^2 f_s d^3v = -\sum_s \frac{1}{r}\left(n_s r^2 \omega_s^2 + K_1 n_s\right),$$
$$\propto -\sum_s \frac{1}{\tilde{r}}\left(\frac{\tilde{r}^2\tilde{\omega}_s^2}{(1+\tilde{r}^2)^2} + \frac{K_1}{(1+\tilde{r}^2)^2}\right),$$

for K_1 a positive constant, and using elementary integrals. The second term on the RHS is seen to cancel with the first term on the RHS of Eq. (5.54), i.e. P_{rr}/r. Also, we see from Eqs. (5.18) and (5.19) that

$$(\boldsymbol{j}\times\boldsymbol{B})_r = -\frac{4k\tau B_0^2}{\mu_0}\frac{\tilde{r}}{(1+\tilde{r}^2)^2},$$

and so we see that the $\boldsymbol{j}\times\boldsymbol{B}$ force has it's origins in $\pi_{\theta\theta}$. Now we are in a position to account for all the terms in force balance,

$$\underbrace{\frac{\partial}{\partial r}\sum_s \frac{n_s}{\beta_s} + \frac{1}{r}\sum_s n_s V_{\theta s}^2 - \frac{1}{r}\sum_s n_s r^2 \omega_s^2}_{\text{"Derivatives" of potentials}} = \underbrace{\sigma E_r - \boldsymbol{\mathcal{F}}_{\mathrm{c}} \cdot \hat{\boldsymbol{e}}_r + (\boldsymbol{j} \times \boldsymbol{B})_r}_{\text{Forces}}.$$

5.5.3.4 Summary of Force Balance Analysis

The conclusions reached from this analysis are somewhat general since some results did not depend on the specific electromagnetic fields $(\boldsymbol{E}, \boldsymbol{B})$. Regardless, we see that

- The electric field sources/balances gradients in the particle number densities
- The centripetal forces are sourced/balanced by the bulk angular flows, $V_{\theta s}(r)$
- The Lorentz force is sourced/balanced by a centripetal-type force, that treats the flow as uniform circular motion, $V_{\theta s} = r\tilde{\omega}_s$, i.e. rotational flows consistent with a rigid-rotor (see Sect. 5.4).

5.5.4 *Plots of the DF*

A characteristic that one immediately looks for in a new DF is the existence of multiple maxima in velocity space, which are a direct indication of non-thermalisation, relevant for the existence of micro-instabilities (e.g. see Gary 2005). Using an analysis very similar to that in Neukirch et al. (2009), we can derive—for a given value of $\tilde{\omega}_s$—conditions on $\tilde{r}$ and either $\tilde{v}_z$ or $\tilde{v}_\theta$, for the existence of multiple maxima in the $\tilde{v}_\theta$ or $\tilde{v}_z$ direction respectively. We present these calculations in Sects. 5.5.4.1 and 5.5.4.2. The most readily understood results are that multiple maxima in the $\tilde{v}_\theta$ direction can only occur for $\tilde{r} > 2/|\tilde{\omega}_s|$, and in the $\tilde{v}_z$ direction for $|\tilde{\omega}_s| > 2$. Given these necessary conditions, one can then calculate that multiple maxima of f_s will occur in the $\tilde{v}_\theta$ direction for $\tilde{v}_z$ bounded above and below, and vice versa.

In Figs. 5.6, 5.7, 5.8 and 5.9 we present plots of the DFs over a range of parameter values. Figures 5.6 and 5.7 show the ion DFs for $k = 0.1$ and $k = 1$ respectively, for all combinations of $\tilde{\omega}_i = 1, 3, \tilde{r} = 0.5, 2$ and $C_s = 0.1, 1$, and with the magnetisation parameter $\delta_i = 1$. As a graphical confirmation of the above discussion, we can only see multiple maxima in the $\tilde{v}_\theta$ direction for $\tilde{r} > 2/|\tilde{\omega}_s|$, and in the $\tilde{v}_z$ direction for $|\tilde{\omega}_s| > 2$, with the appropriate bounds marked by the horizontal/vertical white lines.

Aside from multiple maxima in the orthogonal directions, the DF can also be 'two-peaked'. That is, the DF can have two isolated peaks in $(\tilde{v}_z, \tilde{v}_\theta)$ space. This is seen to occur for Fig. 5.7d, g, h). Hence, f_i is seen to be 'two-peaked' when $k = 1$ for both $\tilde{r} > 2/\tilde{\omega}_i$ and $\tilde{r} < 2/\tilde{\omega}_i$. However, we do not see a two-peaked DF for $k = 0.1$. This seems to suggest that the stronger guide field ($k = 1$) correlates with multiple peaks. Physically, this may correspond to the fact that a homogeneous guide field is consistent with a Maxwellian DF centred on the origin in $(\tilde{v}_z, \tilde{v}_\theta)$ space, given that a Maxwellian contributes zero current. Hence, if the 'main' part/peak of the DF

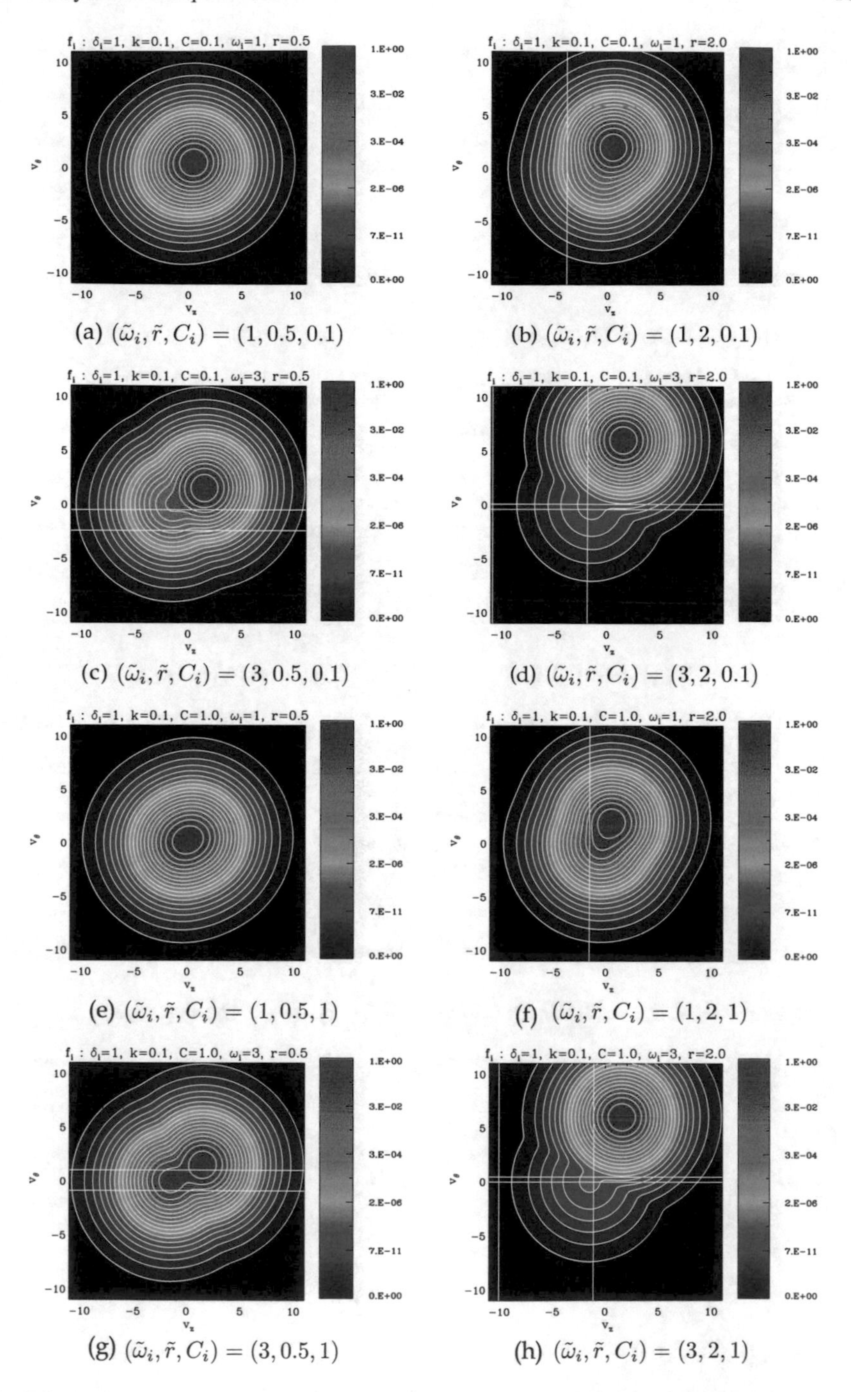

(a) $(\tilde{\omega}_i, \tilde{r}, C_i) = (1, 0.5, 0.1)$

(b) $(\tilde{\omega}_i, \tilde{r}, C_i) = (1, 2, 0.1)$

(c) $(\tilde{\omega}_i, \tilde{r}, C_i) = (3, 0.5, 0.1)$

(d) $(\tilde{\omega}_i, \tilde{r}, C_i) = (3, 2, 0.1)$

(e) $(\tilde{\omega}_i, \tilde{r}, C_i) = (1, 0.5, 1)$

(f) $(\tilde{\omega}_i, \tilde{r}, C_i) = (1, 2, 1)$

(g) $(\tilde{\omega}_i, \tilde{r}, C_i) = (3, 0.5, 1)$

(h) $(\tilde{\omega}_i, \tilde{r}, C_i) = (3, 2, 1)$

Fig. 5.6 Contour plots of the f_i in $(\tilde{v}_z, \tilde{v}_\theta)$ space for an equilibrium with field reversal ($k = 0.1 < 0.5$), for a variety of parameters $(\tilde{\omega}_i, \tilde{r}, C_i)$ and $\delta_i = 1$. The white horizontal/vertical lines indicate the regions in which multiple maxima in either the $\tilde{v}_z$ or $\tilde{v}_z$ directions can occur, if at all. A single line indicates that the 'region' is a line

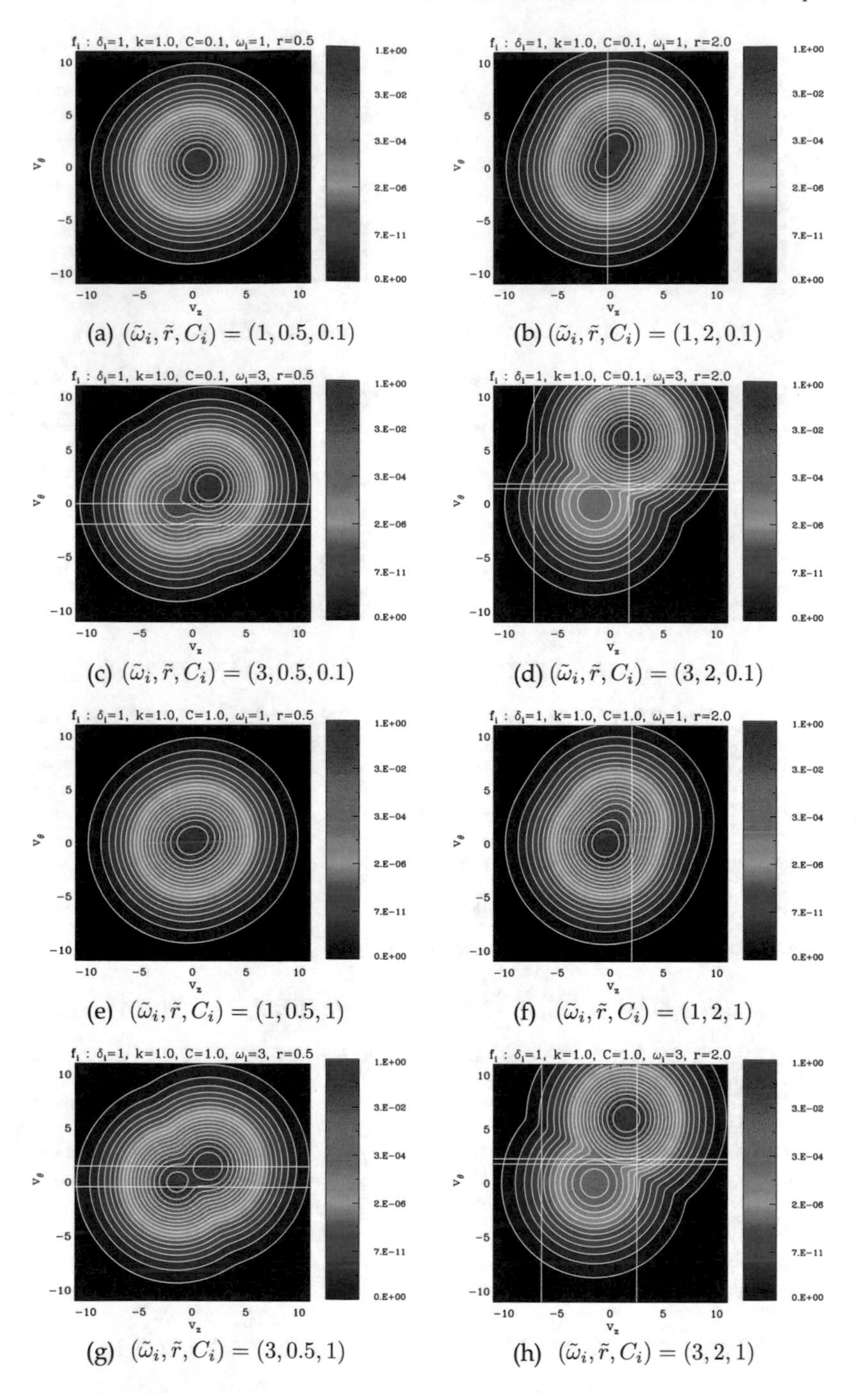

(a) $(\tilde{\omega}_i, \tilde{r}, C_i) = (1, 0.5, 0.1)$
(b) $(\tilde{\omega}_i, \tilde{r}, C_i) = (1, 2, 0.1)$
(c) $(\tilde{\omega}_i, \tilde{r}, C_i) = (3, 0.5, 0.1)$
(d) $(\tilde{\omega}_i, \tilde{r}, C_i) = (3, 2, 0.1)$
(e) $(\tilde{\omega}_i, \tilde{r}, C_i) = (1, 0.5, 1)$
(f) $(\tilde{\omega}_i, \tilde{r}, C_i) = (1, 2, 1)$
(g) $(\tilde{\omega}_i, \tilde{r}, C_i) = (3, 0.5, 1)$
(h) $(\tilde{\omega}_i, \tilde{r}, C_i) = (3, 2, 1)$

Fig. 5.7 Contour plots of f_i in $(\tilde{v}_z, \tilde{v}_\theta)$ space for an equilibrium without field reversal ($k = 1 > 0.5$), for a variety of parameters $(\tilde{\omega}_i, \tilde{r}, C_i)$ and $\delta_i = 1$. The white horizontal/vertical lines indicate the regions in which multiple maxima in either the $\tilde{v}_z$ or $\tilde{v}_z$ directions can occur, if at all. A single line indicates that the 'region' is a line

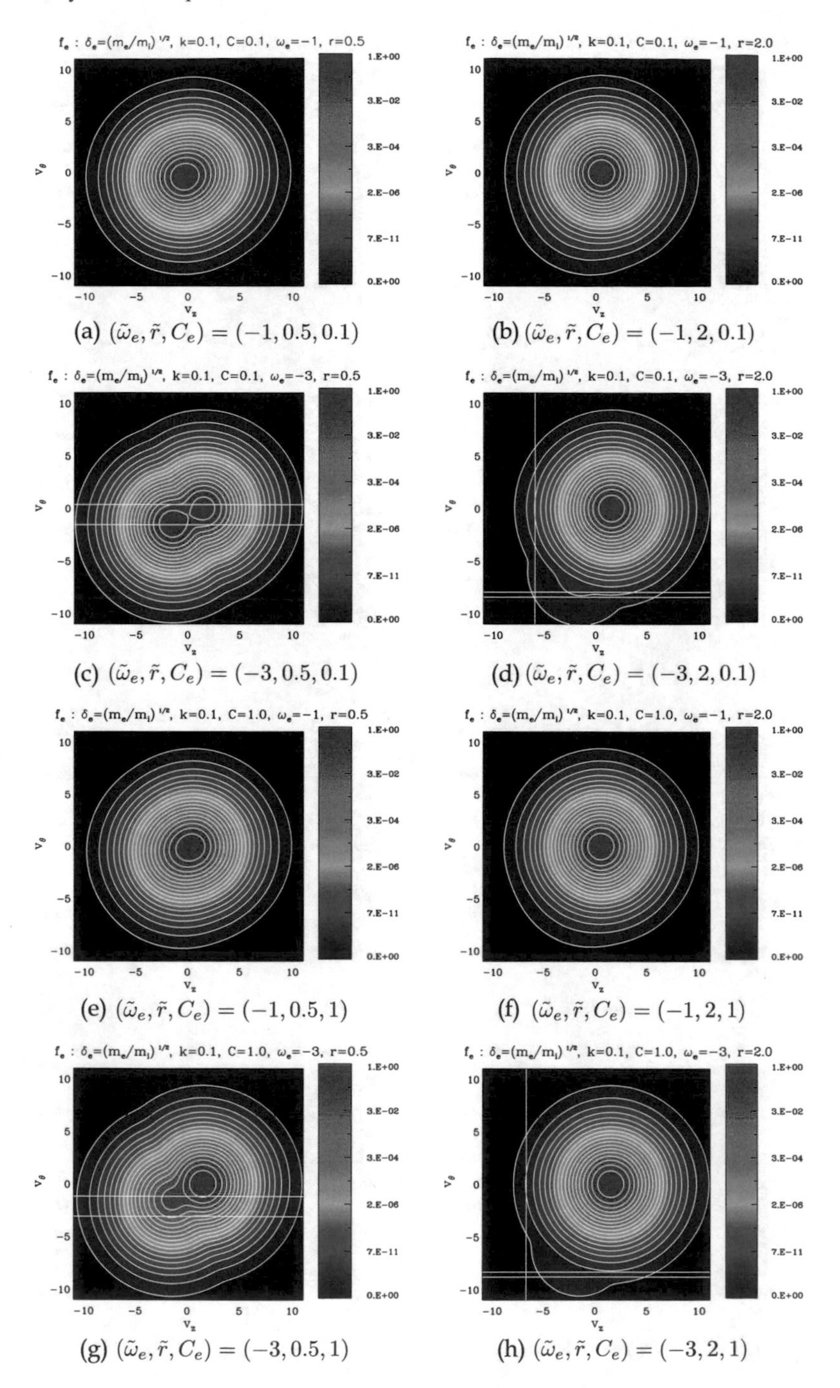

(a) $(\tilde{\omega}_e, \tilde{r}, C_e) = (-1, 0.5, 0.1)$

(b) $(\tilde{\omega}_e, \tilde{r}, C_e) = (-1, 2, 0.1)$

(c) $(\tilde{\omega}_e, \tilde{r}, C_e) = (-3, 0.5, 0.1)$

(d) $(\tilde{\omega}_e, \tilde{r}, C_e) = (-3, 2, 0.1)$

(e) $(\tilde{\omega}_e, \tilde{r}, C_e) = (-1, 0.5, 1)$

(f) $(\tilde{\omega}_e, \tilde{r}, C_e) = (-1, 2, 1)$

(g) $(\tilde{\omega}_e, \tilde{r}, C_e) = (-3, 0.5, 1)$

(h) $(\tilde{\omega}_e, \tilde{r}, C_e) = (-3, 2, 1)$

Fig. 5.8 Contour plots of f_e in $(\tilde{v}_z, \tilde{v}_\theta)$ space for an equilibrium with field reversal ($k = 0.1 < 0.5$), for a variety of parameters $(\tilde{\omega}_e, \tilde{r}, C_e)$ and $\delta_e \approx 1/\sqrt{1836}$. The white horizontal/vertical lines indicate the regions in which multiple maxima in either the $\tilde{v}_z$ or $\tilde{v}_z$ directions can occur, if at all. A single line indicates that the 'region' is a line

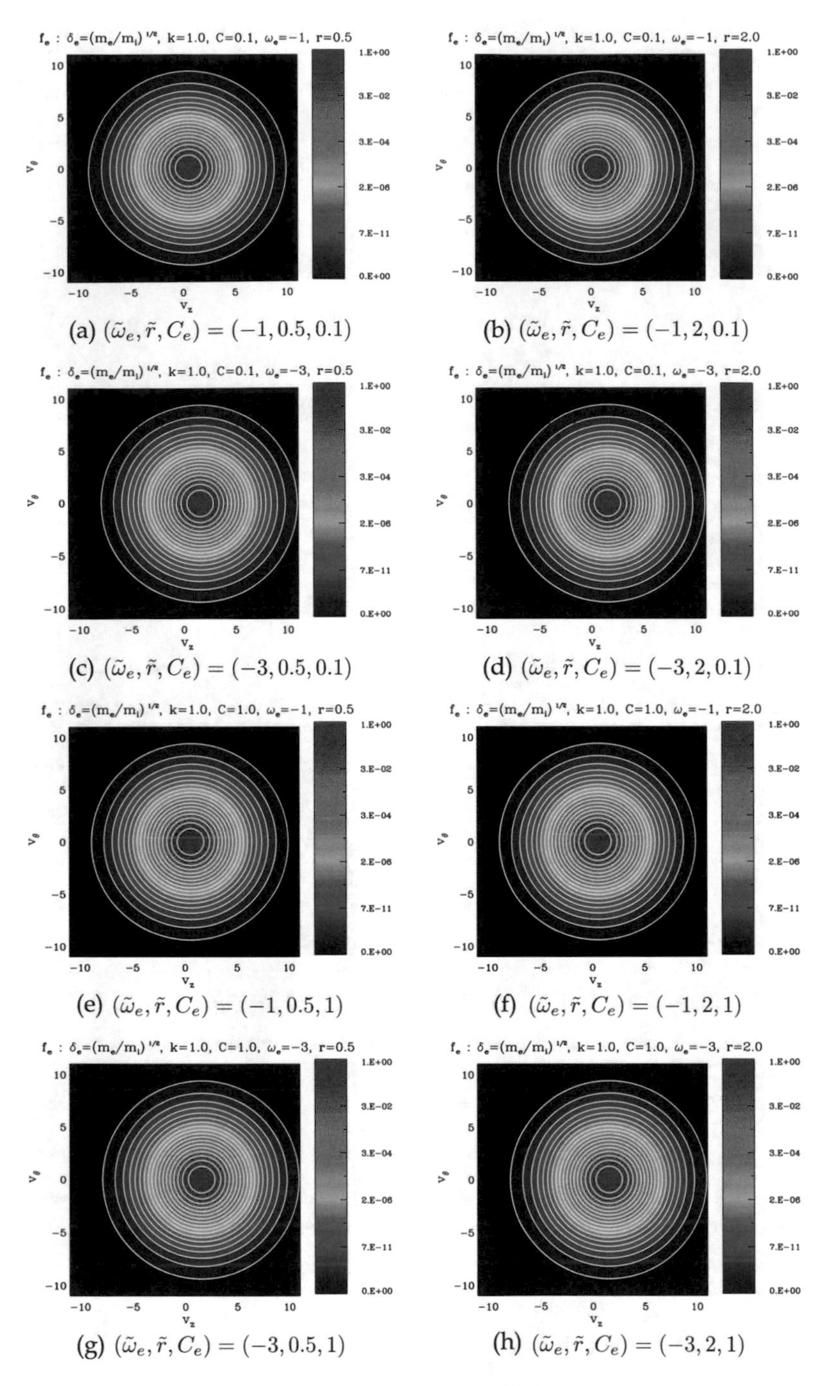

(a) $(\tilde{\omega}_e, \tilde{r}, C_e) = (-1, 0.5, 0.1)$
(b) $(\tilde{\omega}_e, \tilde{r}, C_e) = (-1, 2, 0.1)$
(c) $(\tilde{\omega}_e, \tilde{r}, C_e) = (-3, 0.5, 0.1)$
(d) $(\tilde{\omega}_e, \tilde{r}, C_e) = (-3, 2, 0.1)$
(e) $(\tilde{\omega}_e, \tilde{r}, C_e) = (-1, 0.5, 1)$
(f) $(\tilde{\omega}_e, \tilde{r}, C_e) = (-1, 2, 1)$
(g) $(\tilde{\omega}_e, \tilde{r}, C_e) = (-3, 0.5, 1)$
(h) $(\tilde{\omega}_e, \tilde{r}, C_e) = (-3, 2, 1)$

Fig. 5.9 Contour plots of f_e in $(\tilde{v}_z, \tilde{v}_\theta)$ space for an equilibrium without field reversal ($k = 1 > 0.5$), for a variety of parameters $(\tilde{\omega}_e, \tilde{r}, C_e)$ and $\delta_e \approx 1/\sqrt{1836}$. Note that there are not any multiple maxima in this case

is centred away from the origin, then the Maxwellian contribution from the guide field could contribute a secondary peak. These secondary peaks are seen to be more pronounced when $\tilde{C}_i$ is larger, i.e. the contribution from the second term from the DF is greater.

Figures 5.8 and 5.9 show the electron DFs for $k = 0.1$ and $k = 1$ respectively, for all combinations of $\tilde{\omega}_e = 1, 3$; $\tilde{r} = 0.5, 2$, and $C_e = 0.1, 1$, and with the magnetisation parameter $\delta_e = \delta_i\sqrt{m_e/m_i} \approx 1/\sqrt{1836}$. This choice of magnetisation corresponds to $T_i = T_e$. In general we see DFs with fewer multiple maxima in velocity space than the ion plots, which is physically consistent with the electrons being more magnetised, i.e. more 'fluid-like'. In particular we see no multiple maxima in Fig. 5.9, the case with the stronger background field.

Note that when the electrons have the same magnetisation as the ions, i.e. $\delta_e = \delta_i = 1$, then these marked differences in the velocity-space plots disappear, and we observe a qualitative symmetry $f_i(\tilde{v}_\theta, \tilde{v}_z, r) \propto f_e(-\tilde{v}_\theta, -\tilde{v}_z, r)$.

5.5.4.1 Maxima in v_θ Space

The $\tilde{p}_{rs}$ dependence of the DF is irrelevant to our discussion, and as such can be integrated out. We can also neglect the scalar potential ϕ. The reduced DF, $\tilde{F}_s$, in dimensionless form is

$$\tilde{F}_s = ((\sqrt{2\pi} v_{\mathrm{th},s})^2/n_{0s})\, e^{\tilde{\phi}_s} \int_{-\infty}^{\infty} f_s\, dv_r,$$

which then reads

$$\tilde{F}_s = \exp\left\{-\frac{1}{2}\left[\left(\frac{\tilde{p}_{\theta s}}{\tilde{r}} - \tilde{A}_{\theta s}\right)^2 + \left(\tilde{p}_{zs} - \tilde{A}_{zs}\right)^2\right]\right\} \times$$
$$\left[\exp\left(\tilde{\omega}_s \tilde{p}_{\theta s} + \tilde{U}_{zs}\tilde{P}_{zs}\right) + C_s \exp\left(\tilde{V}_{zs}\tilde{P}_{zs}\right)\right]. \tag{5.55}$$

We have written $\tilde{F}_s$ in terms of the canonical momenta, and so we search for stationary points given by $\partial\tilde{F}_s/\partial\tilde{p}_{\theta s} = 0$, equivalent to $\partial\tilde{F}_s/\partial\tilde{v}_{\theta s} = 0$. Setting $\partial\tilde{F}_s/\partial\tilde{p}_{\theta s} = 0$ gives

$$\begin{aligned}\tilde{p}_{\theta s} - \tilde{r}\tilde{A}_{\theta s} &= \frac{\tilde{\omega}_s \tilde{r}^2}{1 + C_s e^{-\tilde{\omega}_s \tilde{p}_{zs}} e^{-\tilde{\omega}_s \tilde{p}_{\theta s}}} \\ &= \frac{A}{1 + B e^{-\tilde{\omega}_s \tilde{p}_{\theta s}}} := R(\tilde{p}_{\theta s}).\end{aligned} \tag{5.56}$$

To derive a necessary condition for multiple maxima, we analyse the RHS of Eq. (5.56), $R(\tilde{p}_{\theta s})$. This function is bounded between 0 and A, and is monotonically increasing. Hence, using techniques similar to those in Neukirch et al. (2009), a necessary condition for multiple maxima in the DF is that

$$\max_{\tilde{p}_{\theta s}} R'(\tilde{p}_{\theta s}) > 1, \tag{5.57}$$

since the LHS of Eq. (5.56) is a linear function of unit slope in $\tilde{p}_{\theta s}$. This condition can be shown to be equivalent to $A\tilde{\omega}_s/4 > 1$ and so

$$\tilde{\omega}_s^2 > 4\tilde{r}^{-2} \iff \tilde{r} > 2/|\tilde{\omega}_s| \tag{5.58}$$

This demonstrates that for sufficiently small $\tilde{r}$, there cannot exist multiple maxima. Equivalently, this condition will always be satisfied for some $\tilde{r}$, and as such is just a condition on the domain, in $\tilde{r}$, for which multiple maxima can occur. This condition is not sufficient however, as it could still be the case that there exists only one point of intersection (and hence one maximum), depending on the value of B. It is seen that R has unit slope at

$$\tilde{p}_{\theta s}^{\pm} = \frac{1}{\tilde{\omega}_s} \times \left[\ln(2B) - \ln\left(A\tilde{\omega}_s - 2 \pm \sqrt{A\tilde{\omega}_s(A\tilde{\omega}_s - 4)}\right)\right]. \tag{5.59}$$

Clearly R has unit slope for two values of $\tilde{p}_{\theta s}$. After some graphical consideration of the problem, it becomes apparent that B should be bounded above and below for multiple maxima. After elementary consideration of the functional form of (5.56), for example with graph plotting software, we see that multiple maxima in the $\tilde{v}_\theta$ direction can only occur, for a given $\tilde{r}$, when B (and hence $\tilde{v}_z$) satisfies these inequalities for ions

$$\begin{aligned} \tilde{p}_{\theta i}^{+} - R(\tilde{p}_{\theta i}^{+}) - \tilde{r}\tilde{A}_{\theta i} &> 0, \\ \tilde{p}_{\theta i}^{-} - R(\tilde{p}_{\theta i}^{-}) - \tilde{r}\tilde{A}_{\theta i} &< 0, \end{aligned} \tag{5.60}$$

and these for electrons

$$\begin{aligned} \tilde{p}_{\theta e}^{+} - R(\tilde{p}_{\theta e}^{+}) - \tilde{r}\tilde{A}_{\theta e} &< 0, \\ \tilde{p}_{\theta e}^{-} - R(\tilde{p}_{\theta e}^{-}) - \tilde{r}\tilde{A}_{\theta e} &> 0. \end{aligned} \tag{5.61}$$

5.5.4.2 Maxima in v_z Space

We shall once again use the reduced DF defined in Eq. (5.55) in our analysis. Thus, we shall consider $\partial\tilde{F}_s/\partial\tilde{p}_{zs} = 0$, which is equivalent to $\partial\tilde{F}_s/\partial\tilde{v}_{zs} = 0$. Setting $\partial\tilde{F}_s/\partial\tilde{p}_{zs} = 0$ gives

$$
\begin{aligned}
\tilde{p}_{zs} - \tilde{A}_{zs} &= \frac{\tilde{U}_{zs} + C_s \tilde{V}_{zs} e^{-\tilde{\omega}_s(\tilde{p}_{zs}+\tilde{p}_{\theta s})}}{1 + C_s e^{-\tilde{\omega}_s(\tilde{p}_{zs}+\tilde{p}_{\theta s})}} \\
&= \frac{A_1}{1 + B_1 e^{-D_1 \tilde{p}_{zs}}} + \frac{A_2}{1 + B_2 e^{-D_2 \tilde{p}_{zs}}} \\
&:= R_1(\tilde{p}_{zs}) + R_2(\tilde{p}_{zs}) = R(\tilde{p}_{zs}),
\end{aligned}
$$

such that

$$
\begin{aligned}
&A_1 = \tilde{U}_{zs}, \quad A_2 = \tilde{V}_{zs}, \\
&B_1 = C_s e^{-\tilde{\omega}_s \tilde{p}_{\theta s}} = B_2^{-1}, \quad D_1 = \tilde{\omega}_s = -D_2.
\end{aligned}
$$

To derive a necessary condition for multiple maxima, we analyse the RHS of Eq. (5.62). Each R function is bounded and monotonic. Once again using techniques similar to those in Neukirch et al. (2009), a necessary condition for multiple maxima in the DF is that

$$
\max_{\tilde{p}_{zs}} \left(R_1'(\tilde{p}_{zs}) + R_2'(\tilde{p}_{zs}) \right) > 1. \tag{5.62}
$$

After some algebra this condition can be shown to be equivalent to $\tilde{\omega}_s^2/4 > 1$ and so

$$
|\tilde{\omega}_s| > 2. \tag{5.63}
$$

This condition is not sufficient however, as it could still be the case that there exists only one point of intersection, depending on the value of $B_1 (= 1/B_2)$. The transition between 3 points of intersection and one occurs at the value of B_1 for which the straight line of slope unity through $\tilde{p}_{zs} = 0$ just touches $R_1(\tilde{p}_{zs}) + R_2(\tilde{p}_{zs})$ at the point where it also has unit slope. It is readily seen that $R_1 + R_2$ has unit slope at

$$
\tilde{p}_{zs}^{\pm} = \frac{1}{\tilde{\omega}_s} \left[\ln(2B_1) - \ln\left(\tilde{\omega}_s^2 - 2 \pm \sqrt{\tilde{\omega}_s^2(\tilde{\omega}_s^2 - 4)} \right) \right]. \tag{5.64}
$$

Clearly R has unit slope for two values of $\tilde{p}_{zs}$. Once again, after some graphical consideration of the problem, it becomes apparent that B_1 should be bounded above and below for multiple maxima. After elementary consideration of the functional form of (5.62), for example with graph plotting software we see that multiple maxima in the $\tilde{v}_z$ direction can only occur, for a given $\tilde{r}$, when B_1 (and hence $\tilde{v}_\theta$) satisfies these inequalities for ions

$$
\begin{aligned}
&\tilde{p}_{zi}^{+} - R(\tilde{p}_{zi}^{+}) - \tilde{A}_{zi} > 0, \\
&\tilde{p}_{zi}^{-} - R(\tilde{p}_{zi}^{-}) - \tilde{A}_{zi} < 0,
\end{aligned} \tag{5.65}
$$

and these for electrons

$$\tilde{p}_{ze}^{+} - R(\tilde{p}_{ze}^{+}) - \tilde{A}_{ze} < 0,$$
$$\tilde{p}_{ze}^{-} - R(\tilde{p}_{ze}^{-}) - \tilde{A}_{ze} > 0. \quad (5.66)$$

5.6 Summary

In this chapter we have calculated 1D collisionless equilibria for a continuum of magnetic field models based on the GH flux tube, with an additional constant background field in the axial direction. This study was motivated by a desire to extend the existing methods for solutions of the 'inverse problem in Vlasov equilibria' in Cartesian geometry, to cylindrical geometry.

In Sect. 5.3.3 we calculated the fluid equations of motion for a 1D system with azimuthal and axial flows, found by taking the first order velocity moment of the Vlasov equation in cylindrical coordinates. The presence of centripetal forces in the equation of motion demonstrated that it may be difficult to find Vlasov equilibrium DFs self-consistent with force-free fields.

However, initial efforts focussed on solving for the exact force-free GH field, but this seems impossible due to the centripetal forces, and this conclusion is somewhat corroborated by Vinogradov et al. (2016). The GH field in particular was chosen as it represents the 'natural' analogue of the Force-Free Harris Sheet in cylindrical geometry, a magnetic field whose VM equilibria have been the subject of recent study, (Harrison and Neukirch 2009a; Neukirch et al. 2009; Wilson and Neukirch 2011; Abraham-Shrauner 2013; Kolotkov et al. 2015), as well as the work detailed in Chaps. 2 and 3, featuring work from Allanson et al. (2015, 2016)

A background field was introduced, and an equilibrium DF was found that reproduces the required magnetic field, i.e. solves Ampère's Law. It is the presence of the background field that allows us to solve Vlasov's equation and Ampère's Law, and it appears physically necessary as it introduces an 'asymmetry'; namely an extra term into the equation of motion whose sign depends explicitly on species. In contrast to the 'demands' of insisting on a particular magnetic field, no condition was made on the electric field. The DF allows both electrically neutral and non-neutral configurations, and in the case of non-neutrality we find an exact and explicit solution to Poisson's equation for an electric field that decays like $1/r$ far from the axis. We note here that the type of solutions derived in this chapter could—after a Galilean transformation—be interpreted as 1D BGK modes with finite magnetic field (see Abraham-Shrauner 1968; Ng and Bhattacharjee 2005; Grabbe 2005; Ng et al. 2006 for example, to provide some context).

An analysis of the physical properties of the DF was given in Sect. 5.5, with some particularly detailed calculations in Sects. 5.5.4.1 and 5.5.4.2. The dependence of the sign of the charge density (and hence the electric field) on the bulk ion and electron rotational flows was analysed, with a physical interpretation given. Essentially the argument states that the electric field exists in order to balance the difference in the centrifugal forces (in the co-moving frame) between the two species. The DF was found to be able to give sub-unity values of the plasma beta, should this be required/desirable given the relevant physical system that it is intended to model. In Sect. 5.5.3 we performed a detailed analysis of the relationship between individual terms in the equation of motion. For clarity, the conclusions drawn for the macroscopic equilibrium considered in this chapter are that the electric field sources/balances gradients in the particle number densities; the centripetal forces are sourced/balanced by the bulk angular flows; and the $\boldsymbol{j} \times \boldsymbol{B}$ force is sourced/balanced by a centripetal-type force, that treats the flow as uniform circular motion, i.e. rotational flows consistent with a rigid-rotor (see Sect. 5.4). The final part of the analysis focussed on plotting the DF in velocity space, for certain parameter values, and at different radii. Mathematical conditions were found that determine whether or not the DF could have multiple maxima in the orthogonal directions in velocity space, and these are corroborated by the plots of the DFs. For certain parameter values, the DF was also seen to have two separate, isolated peaks. This non-thermalisation suggests the existence of microinstabilities, for a certain choice of parameters.

Further work could involve a deeper analysis of the properties of the DFs and their stability. This work has also raised a fundamental question: 'is it possible to describe a 1D force-free collisionless equilibrium in cylindrical geometry?' Preliminary investigations seem to suggest that it is not possible. It would also be of value to find out whether the relationships derived between individual terms in the equation of motion are totally general in nature, and if not, to what extent do they apply?

References

B. Abraham-Shrauner, Exact, stationary wave solutions of the nonlinear Vlasov equation. Phys. Fluids **11**, 1162–1167 (1968)

B. Abraham-Shrauner, Force-free Jacobian equilibria for Vlasov-Maxwell plasmas. Phys. Plasmas **20**(10), 102117 (2013)

O. Allanson, T. Neukirch, S. Troscheit, F. Wilson, From onedimensional fields to Vlasov equilibria: theory and application of Hermite polynomials. J. Plasma Phys. **82**(3), 905820306 (2016)

O. Allanson, T. Neukirch, F.Wilson, S. Troscheit, An exact collisionless equilibrium for the Force-Free Harris sheet with low plasma beta. Phys. Plasmas **22**(10), 102116 (2015)

O. Allanson, F. Wilson, T. Neukirch, Neutral and non-neutral collisionless plasma equilibria for twisted flux tubes: the Gold-Hoyle model in a background field. Phys. Plasmas **23**(9), 092106 (2016)

W. Alpers, Steady state charge neutral models of the magnetopause. Astrophys. Space Sci. **5**, 425–437 (1969)

A.V. Artemyev, A model of one-dimensional current sheet with parallel currents and normal component of magnetic field. Phys. Plasmas **18**(2), 022104 (2011)

D.B. Batchelor, R.C. Davidson, Kinetic description of linear thetapinch equilibria. J. Plasma Phys. **14**, 77–92 (1975)

W.H. Bennett, Magnetically self-focussing streams. Phys. Rev. **45**(12), 890–897 (1934)

J. Birn, E. Priest, Reconnection of magnetic fields: magnetohydrodynamics and collisionless theory and observations. Cambridge University Press (2007)

N.A. Bobrova, S.I. Syrovatskiĭ, Violent instability of one-dimensional forceless magnetic field in a rarefied plasma. Sov. J. Exp. Theor. Phys. Lett. **30**, 535-+ (1979)

N.A. Bobrova, S.V. Bulanov, J.I. Sakai, D. Sugiyama, Force-free equilibria and reconnection of the magnetic field lines in collisionless plasma configurations. Phys. Plasmas **8**, 759–768 (2001)

A.L. Borg, M.G.G.T. Taylor, J.P. Eastwood, Observations of magnetic flux ropes during magnetic reconnection in the Earth's magnetotail. Annales Geophysicae **30**, 761–773 (2012)

J.E. Borovsky, Flux tube texture of the solar wind: strands of the magnetic carpet at 1 AU? J. Geophys. Res. (Space Phys.) **113**, A08110 (2008)

A. Bottino, A.G. Peeters, R. Hatzky, S. Jolliet, B.F. McMillan, T.M. Tran, L. Villard, Nonlinear low noise particle-in-cell simulations of electron temperature gradient driven turbulence. Phys. Plasmas **14**(1), 010701 (2007)

P. Carlqvist, Cosmic electric currents and the generalized Bennett relation. Astrophys. Space Sci. **144**, 73–84 (1988)

P.J. Channell, Exact Vlasov-Maxwell equilibria with sheared magnetic fields. Phys. Fluids **19**, 1541–1545 (1976)

S.W. Channon, M. Coppins, Z-pinch Vlasov-fluid equilibria including sheared-axial flow. J. Plasma Phys. **66**, 337–347 (2001)

S.W.H. Cowley, C.J. Owen, A simple illustrative model of open flux tube motion over the dayside magnetopause. Planet. Space Sci. **37**, 1461–1475 (1989)

S.C. Cowley, B. Cowley, S.A. Henneberg, H.R. Wilson, Explosive instability and erupting flux tubes in a magnetized plasma. Proc. R. Soc. Lond. Ser. A **471**, 20140913 (2015)

R.C. Davidson. *Physics of Nonneutral Plasmas* (World Scientific Press, 2001)

G.A. DiBraccio, J.A. Slavin, S.M. Imber, D.J. Gershman, J.M. Raines, C.M. Jackman, S.A. Boardsen, B.J. Anderson, H. Korth, T.H. Zurbuchen, R.L. McNutt, S.C. Solomon, MESSENGER observations of flux ropes in Mercury's magnetotail. Planet. Space Sci. **115**, 77–89 (2015)

J.P. Eastwood, T.D. Phan, P.A. Cassak, D.J. Gershman, C. Haggerty, K. Malakit, M.A. Shay, R. Mistry, M. Øieroset, C.T. Russell, J.A. Slavin, M.R. Argall, L.A. Avanov, J.L. Burch, L.J. Chen, J.C. Dorelli, R.E. Ergun, B.L. Giles, Y. Khotyaintsev, B. Lavraud, P.A. Lindqvist, T.E. Moore, R. Nakamura, W. Paterson, C. Pollock, R.J. Strangeway, R.B. Torbert, S. Wang, Ion-scale secondary flux ropes generated by magnetopause reconnection as resolved by MMS. Geophys. Res. Lett. **43**, 4716–4724 (2016)

A. El-Nadi, G. Hasselberg, A. Rogister, A p_θ, p_z dependent equilibrium solution for the cylindrical pinch. Phys. Lett. A **56**, 297–298 (1976)

D.F. Escande, *What is a Reversed Field Pinch?* (World Scientific, 2015), pp. 247–286. (Chap. 9)

Y. Fan, Magnetic fields in the solar convection zone. Living Rev. Solar Phys. **6**(4) (2009)

R. Fitzpatrick, *Plasma Physics: An Introduction* (CRC Press, Taylor & Francis Group, 2014)

J.P. Freidberg, *Ideal Magnetohydrodynamics* (Plenum Publishing Corportation, 1987)

S.P. Gary, *Theory of Space Plasma Microinstabilities*, Cambridge Atmospheric and Space Science Series (Cambridge University Press, 2005)

T. Gold, F. Hoyle, On the origin of solar flares. Mon. Not. R. Astrono. Soc. **120**, 89 (1960)

C.L. Grabbe, Trapped-electron solitary wave structures in a magnetized plasma. Phys. Plasmas **12**(7), 072311 (2005)

P. Gratreau, P. Giupponi, Vlasov equilibria of cylindrical relativistic electron beams of arbitrary high intensity. Phys. Fluids **20**, 487–493 (1977)

J.M. Greene, One-dimensional Vlasov-Maxwell equilibria. Phys. Fluids B **5**, 1715–1722 (1993)

C.J. Ham, S.C. Cowley, G. Brochard, H.R. Wilson, Nonlinear stability and saturation of ballooning modes in tokamaks*". Phys. Rev. Lett. **116**(23), 235001 (2016)

D.A. Hammer, N. Rostoker, Propagation of high current relativistic electron beams. Phys. Fluids **13**, 1831–1850 (1970)

E.G. Harris, On a plasma sheath separating regions of oppositely directed magnetic field. Nuovo Cimento **23**, 115 (1962)

M.G. Harrison, T. Neukirch, One-dimensional Vlasov-Maxwell equilibrium for the force-free Harris sheet. Phys. Rev. Lett. **102**(13), pp. 135003-+ (2009a)

M.G. Harrison, T. Neukirch, Some remarks on one-dimensional forcefree Vlasov-Maxwell equilibria. Phys. Plasmas **16**(2), 022106-+ (2009b)

A.W. Hood, P.K. Browning, R.A.M. van der Linden, Coronal heating by magnetic reconnection in loops with zero net current. Astron. Astrophys. **506**, 913–925 (2009)

A.W. Hood, P.J. Cargill, P.K. Browning, K.V. Tam, An MHD avalanche in a multi-threaded coronal loop. Astrophys. J. **817**(5), 5 (2016)

J.D. Huba, *NRL PLASMA FORMULARY Supported by The Office of Naval Research* (Naval Research Laboratory, Washington, DC, 2013), pp. 1–71

K.K. Khurana, M.G. Kivelson, L.A. Frank, W.R. Paterson, Observations of magnetic flux ropes and associated currents in Earth's magnetotail with the Galileo spacecraft. Geophys. Res. Lett. **22**, 2087–2090 (1995)

M.G. Kivelson, K.K. Khurana, Models of flux ropes embedded in a Harris neutral sheet: force-free solutions in low and high beta plasmas. J. Geophys. Res. **100**, 23637–23646 (1995)

D.Y. Kolotkov, I.Y. Vasko, V.M. Nakariakov, Kinetic model of forcefree current sheets with non-uniform temperature. Phys. Plasmas **22**(11), 112902 (2015)

N.N. Komarov, N.M. Fadeev, Plasma in a self-consistent magnetic field. Sov. Phys. JETP **14**(2), 528–533 (1962)

L.D. Landau, E.M. Lifshitz, *The Classical Theory of Fields*, Course of Theoretical Physics (Elsevier Science, 2013)

U. Leonhardt, T. Philbin, *Geometry and Light: The Science of Invisibility*, Dover Books on Physics (Dover Publications, 2012)

H. Li, G. Lapenta, J.M. Finn, S. Li, S.A. Colgate, Modeling the large-scale structures of astrophysical jets in the magnetically dominated limit. Astrophys. J. **643**, 92–100 (2006)

T. Magara, D.W. Longcope, Injection of magnetic energy and magnetic helicity into the solar atmosphere by an emerging magnetic flux tube. Astrophys. J. **586**, 630–649 (2003)

S.M. Mahajan, Exact and almost exact solutions to the Vlasov-Maxwell system. Phys. Fluids B **1**, 43–54 (1989)

E. Marsch, Kinetic Physics of the Solar Corona and Solar Wind. Living Rev. Sol. Phys. **3**(1) (2006). http://www.livingreviews.org/lrsp-2006-1

G.E. Marsh, *Force-Free Magnetic Fields: Solutions, Topology and Applications*. (World Scientific, Singapore, 1996)

A.I. Morozov, L.S. Solov'ev, A kinetic examination of some equilibrium plasma configurations. Sov. Phys. JETP **13**, 927–932 (1961)

H.E. Mynick, W.M. Sharp, A.N. Kaufman, Realistic Vlasov slab equilibria with magnetic shear. Phys. Fluids **22**, 1478–1484 (1979)

T. Neukirch, F. Wilson, M.G. Harrison, A detailed investigation of the properties of a Vlasov-Maxwell equilibrium for the force-free Harris sheet. Phys. Plasmas **16**(12), 122102 (2009)

W.A. Newcomb, Hydromagnetic stability of a diffuse linear pinch. Ann. Phys. **10**, 232–267 (1960)

C.S. Ng, A. Bhattacharjee, Bernstein-Greene-Kruskal modes in a three- dimensional plasma. Phys. Rev. Lett. **95**(24), 245004 (2005)

C.S. Ng, A. Bhattacharjee, F. Skiff, Weakly collisional Landau damping and three-dimensional Bernstein-Greene-Kruskal modes: new results on old problems. Phys. Plasmas **13**(5), 055903 (2006)

R.B. Nicholson, Solution of the Vlasov equations for a plasma in an externally uniform magnetic field. Phys. Fluids **6**, 1581–1586 (1963)

D. Pfirsch, Mikroinstabilitäten vom Spiegeltyp in inhomogenen Plasmen. Zeitschrift Naturforschung Teil A **17**, 861–870 (1962)

D.H. Pontius Jr., R.A. Wolf, Transient flux tubes in the terrestrial magnetosphere. Geophys. Res. Lett. **17**, 49–52 (1990)

E. Priest, *Magnetohydrodynamics of the Sun* (2014)

E.R. Priest, J.F. Heyvaerts, A.M. Title, A flux-tube tectonics model for solar coronal heating driven by the magnetic carpet. Astrophys. J. **576**, 533–551 (2002)

A.D. Rogava, S. Poedts, S.M. Mahajan, Shear-driven wave oscillations in astrophysical flux tubes. Astron. Astrophys. **354**, 749–759 (2000)

L.I. Rudakov, A.L. Velikovich, J. Davis, J.W. Thornhill, J.L. Giuliani Jr., C. Deeney, Buoyant magnetic flux tubes enhance radiation in Z pinches. Phys. Rev. Lett. **84**, 3326–3329 (2000)

C.T. Russell, R.C. Elphic, Observation of magnetic flux ropes in the Venus ionosphere. Nature **279**, 616–618 (1979)

F. Santini, H. Tasso, Vlasov equation in orthogonal co-ordinates. 1970 Internal Report IC/70/49 (1970)

T. Sato, M. Tanaka, T. Hayashi, T. Shimada, K. Watanabe, Formation of field twisting flux tubes on the magnetopause and solar wind particle entry into the magnetosphere. Geophys. Res. Lett. **13**, 801–804 (1986)

K. Schindler, *Physics of Space Plasma Activity* (Cambridge University Press, 2007)

A. Sestero, Self-consistent description of a warm stationary plasma in a uniformly sheared magnetic field. Phys. Fluids **10**, 193–197 (1967)

J.A. Slavin, R.P. Lepping, J. Gjerloev, D.H. Fairfield, M. Hesse, C.J. Owen, M.B. Moldwin, T. Nagai, A. Ieda, T. Mukai, Geotail observations of magnetic flux ropes in the plasma sheet. J. Geophys. Res. (Space Phys.) **108**, 1015 (2003)

E. Tassi, F. Pegoraro, G. Cicogna, Solutions and symmetries of forcefree magnetic fields. Phys. Plasmas **15**(9), 092113-+ (2008)

H. Tasso, G.N. Throumoulopoulos, FAST TRACK COMMUNICATION: On the Vlasov approach to tokamak equilibria with flow. J. Phys. A Math. Gen. **40**, 631 (2007)

H. Tasso, G. Throumoulopoulos, Tokamak-like Vlasov equilibria. Eur. Phys. J. D **68**(175), 175 (2014)

V.S. Titov, K. Galsgaard, T. Neukirch, Magnetic pinching of hyperbolic flux tubes I. Basic estimations. Astrophys. J. **582**, 1172–1189 (2003)

T. Török, B. Kliem, The evolution of twisting coronal magnetic flux tubes. Astron. Astrophys. **406**, 1043–1059 (2003)

H.S. Uhm, R.C. Davidson, Kinetic equilibrium properties of relativistic non-neutral electron flow in a cylindrical diode with applied magnetic field. Phys. Rev. A **31**, 2556–2569 (1985)

A.A. Vinogradov, I.Y. Vasko, A.V. Artemyev, E.V. Yushkov, A.A. Petrukovich, L.M. Zelenyi, Kinetic models of magnetic flux ropes observed in the Earth magnetosphere. Phys. Plasmas **23**(7), 072901 (2016)

Y.-M. Wang, N.R. Sheeley Jr., Solar wind speed and coronal flux-tube expansion. Astrophys. J. **355**, 726–732 (1990)

T. Wiegelmann, T. Sakurai, Solar Force-free Magnetic Fields. Living Rev. Sol. Phys. **9**(5) (2012)

F. Wilson, T. Neukirch, A family of one-dimensional Vlasov-Maxwell equilibria for the force-free Harris sheet. Phys. Plasmas **18**(8), 082108 (2011)

Y.Y. Yang, C. Shen, Y.C. Zhang, Z.J. Rong, X. Li, M. Dunlop, Y.H. Ma, Z.X. Liu, C.M. Carr, H. Rème, The force-free configuration of flux ropes in geomagnetotail: cluster observations. J. Geophys. Res. (Space Phys.) **119**, 6327–6341 (2014)

Chapter 6
Discussion

For God's sake, stop researching for a while and begin to think.
Walter Hamilton Moberley

The details of the main results of this thesis have been explained in the preambles and summaries of Chaps. 2, 3, 4 and 5, and as such we shall not duplicate that information. Here, it is the intention to place the motivation of the work and the results in context with regards to personal research direction, broader questions, and suggestions for future work.

6.1 Context

The overarching physical motivation for the work in this thesis is perhaps embodied by—and has its roots in—the 'GEM challenge': '*The goal is to identify the essential physics which is required to model collisionless magnetic reconnection*', (Birn et al. 2001). However, this thesis does not focus on the analysis of instability and reconnection itself. The results in this thesis are on the theoretical modelling of Vlasov-Maxwell equilibria, with the approach being a mixture of 'general scientific curiosity' (e.g. Chaps. 2 and 5), and the application to particular physical problems (e.g. Chaps. 3, 4 and 5).

6.1.1 Current Sheets

Much of the research effort in tackling the GEM challenge has been spent on antiparallel (i.e. $B_x(z) = -B_x(-z)$) reconnection, with initial equilibrium conditions as

O. Allanson, *Theory of One-Dimensional Vlasov-Maxwell Equilibria*,
Springer Theses, https://doi.org/10.1007/978-3-319-97541-2_6

symmetric 1D current sheets (e.g. see Hesse et al. 2001; Birn et al. 2005 for examples with and without guide fields B_y respectively). In particular, the Harris current sheet model (or some modification) is very frequently used, in no small part due to the well-known exact Vlasov-Maxwell equilibrium DF (Harris 1962),

$$f_s = \frac{n_{0s}}{(\sqrt{2\pi}\, v_{\mathrm{th},s})^3} e^{-\beta_s(H_s - u_{ys} p_{ys})}.$$

It is possible to approximate force-free ($\boldsymbol{j} \times \boldsymbol{B} = \boldsymbol{0}$) conditions, relevant to the $\beta_{pl} \ll 1$ conditions in the solar corona, by assuming a strong, uniform guide field $B_y(z) = B_{y0} \gg B_{x0}$,

$$\boldsymbol{B} = (B_{x0} \tanh \tilde{z},\, B_{y0},\, 0).$$

However, as discussed in Chap. 3, the nature of such an equilibrium does not accurately represent a true force-free equilibrium, such as the force-free Harris sheet,

$$\boldsymbol{B} = B_0(\tanh \tilde{z},\, \mathrm{sech}\tilde{z},\, 0).$$

Until the discovery of the first VM equilibrium DF for a nonlinear force-free field (the *Harrison-Neukirch* equilibrium for the force-free Harris sheet) by Harrison and Neukirch (2009a), the analysis of reconnection and instability of force-free fields had to be limited to the use of exact initial conditions for a uniform strong guide field configuration, e.g. Ricci et al. (2004); the use of inexact initial conditions (drifting Maxwellians) for an exact nonlinear force-free field (e.g. Birn and Hesse 2010); or one would have to use a linear force-free model (e.g. Bobrova et al. 2001), for which one cannot isolate and study a single current sheet. We are now beginning to see the first analyses of linear stability (Wilson et al. 2017), and reconnection (Wilson et al. 2016) for exact nonlinear force-free current sheet models.

The Harrison-Neukirch equilibrium does have one fairly significant drawback, with regards to its use in a low plasma beta environment. Due to technical reasons regarding the manner in which the Vlasov-Maxwell equilibrium was constructed, β_{pl} is bounded below by unity. This feature motivated our investigations of low-beta Vlasov-Maxwell equilibria for the force-free Harris sheet (Allanson et al. 2015, 2016), as discussed in Chap. 3. The key step in reducing the lower bound for β_{pl}, was the use of pressure tensor transformation techniques, as discussed in Harrison and Neukirch (2009b), and for which we chose an exponential function. This transformation made the inverse problem (Channell 1976) difficult to solve, and confidence in the solution necessitated some rigorous mathematical work (see Allanson et al. 2016) and Chap. 2.

It is now established that '*magnetic reconnection relies on the presence of a diffusion region, where collisionless or collisional plasma processes facilitate the changes in magnetic connection through the generation of dissipative electric fields*' (Hesse et al. 2011). The very recent (and current) NASA MMS mission is able to make in-situ diffusion region measurements on kinetic scales for the very first time (Burch et al. 2016; Hesse et al. 2016). The satellite will focus on the dayside magnetopause in the first phase of its mission, and the magnetotail in the second phase. Current sheets in the dayside magnetopause are typically of a rather different nature than those of the symmetric Harris sheet type, by virtue of the asymmetric conditions either side of the current sheet. The magnetosheath side is characterised by an enhanced thermal pressure and depleted magnetic pressure, and vice versa for the magnetosphere side. Exact analytical (Alpers 1969) and numerical (Belmont et al. 2012; Dorville et al. 2015) Vlasov-Maxwell equilibria are few in number, and so the work in Chap. 4 and Allanson et al. (2017) is targeted towards improving this situation. In particular, the exact analytical solution due to Alpers (1969) has different bulk flow properties to the one that we present.

6.1.2 Flux Tubes

Localised currents need not always obey a planar geometry; flux tubes play an important role in confinement and subsequent energy release in many areas of plasma physics (see Chap. 5), and particularly in the solar corona (e.g. see Wiegelmann and Sakurai 2012; Hood et al. 2016), as well as the extended structure of magnetic islands, perpendicular to current sheets in the magnetopause and magnetotail (e.g. see Kivelson and Khurana 1995; Vinogradov et al. 2016). Hence it was with a combination of mathematical curiosity, and a desire to model nonlinear force-free flux tubes, that we attempted to calculate exact Vlasov-Maxwell equilibria for the Gold-Hoyle flux tube (Gold and Hoyle 1960), the natural analogue of the force-free Harris sheet in cylindrical geometry (Tassi et al. 2008). The work is detailed in Chap. 5 and Allanson et al. (2016), and in fact we were unable to find solutions for the exact nonlinear force-free Gold-Hoyle model. However, the magnetic field can be arbitrarily close to a force-free field if desired. An interesting feature of the analysis focussed on the need to include non-neutrality and non-zero electric fields in the equilibrium, brought about by charge separation effects, inherent in the rotational motion of particles with different masses.

6.2 Broader Theoretical Questions

6.2.1 The Pressure Tensor

In a one-dimensional and z-dependent geometry, the 'keystone' of the inverse problem is the pressure tensor component $P_{zz}(A_x, A_y)$: given a magnetic field, one first attempts to calculate P_{zz}, and then self consistent distribution functions. The main theoretical/mathematical developments in this thesis (related to Cartesian geometry) have focussed on the second step in this process, i.e. calculating self-consistent DFs, of the form

$$f_s = \frac{n_{0s}}{(\sqrt{2\pi}\,v_{\text{th},s})^3} e^{-\beta_s H_s} g_s(p_{xs}, p_{ys}),$$

given a $P_{zz}(A_x, A_y)$. However, there remain important questions about the determination of the P_{zz} function itself.

As discussed in Chaps. 1 and 4, the problem of determining $P_{zz}(A_x, A_y)$ given a magnetic field (in force balance) is analogous to that of determining the shape of a conservative potential function, $\mathcal{V}(\boldsymbol{x})$, given the knowledge of the particle trajectory, $\boldsymbol{x}(t)$, and the value of the potential along the trajectory, $\mathcal{V}(t)$. In the case of 1D force-free fields there is an algorithmic path that determines a valid form of P_{zz} (e.g. see Chap. 3, Harrison and Neukirch 2009b). The question remains: 'to what extent is it possible to find self-consistent P_{zz} functions for a given magnetic field, and what are they?'

One other feature of interest is the solubility of Ampère's Law,

$$\frac{\partial P_{zz}}{\partial \boldsymbol{A}} = -\frac{1}{\mu_0}\frac{d^2 \boldsymbol{A}}{dz^2},$$

with respect to different P_{zz} expressions. As demonstrated in Chap. 3 and Harrison and Neukirch (2009b) for the case of force-free fields; given one P_{zz} that satisfies Ampère's Law, there exist infinitely many others. There are two obvious questions here. Firstly, it would be interesting to investigate if there are ways to transform the Harrison-Neukirch pressure function to allow sub-unity values of the plasma beta, in a way that is more readily soluble and easier to manipulate numerically than the result found in Chap. 3 and Allanson et al. (2015, 2016). Secondly, is it in any way possible to extend the pressure transformation theory for force-free equilibria to non force-free equilibria? If so, then the theory is to be expected to be more complicated than for force-free fields, which relies on P_{zz} being a constant when evaluated along the force-free trajectory $(A_x(z), A_y(z))$.

6.2.2 Non-uniqueness

One clear challenge is to marry together the need for individual, exact solutions of the inverse problem for Vlasov-Maxwell equilibria, versus the fact that there are in principle infinitely many solutions. In essence, how do we know that a given Vlasov-Maxwell equilibrium is appropriate physically? In Chap. 1 we gave arguments for suggesting why distribution functions of the form in Sect. 6.2.1 were reasonable on both physical and mathematical grounds. In particular, this form of distribution function bears a strong resemblance to a (drifting) Maxwellian. Hence, provided the g_s function is not too 'exotic', it seems reasonable that these distribution functions can—for a certain choice of microscopic parameters—minimise the free energy

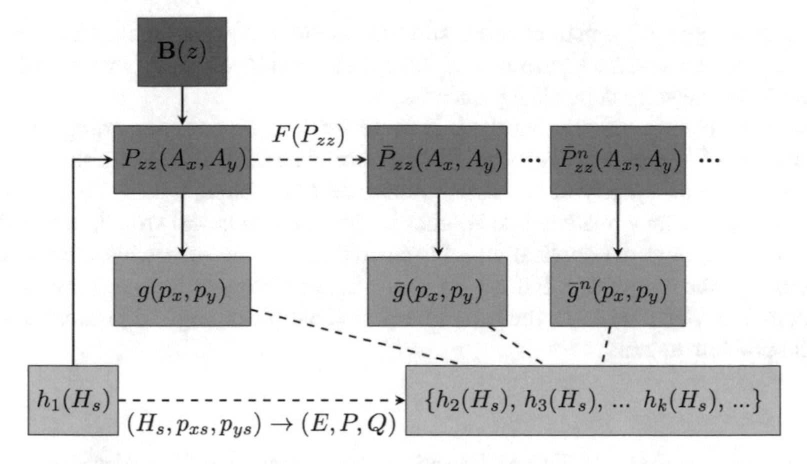

Fig. 6.1 A schematic representation of the inverse problem in Vlasov-Maxwell equilibria

(maximise the entropy) in a plasma, given certain constraints such as the conservation of energy in a closed system (e.g. see Schindler 2007).

The inverse problem is characterised by non-uniqueness on the level of the P_{zz} for a given $\boldsymbol{B}$, and on the level of f_s for a given P_{zz}. It would be of interest to see if—given a distribution function of the form in Sect. 6.2.1—the inversion of the Weierstrass transform gives a unique solution and if not, whether the inversion method (e.g. Fourier transform or Hermite polynomial expansion) has an effect on the outcome. As discussed in Chap. 2, these considerations are related to the 'backwards uniqueness of the heat equation' (Evans 2010), with g_s and P_{zz} somewhat equivalent to the initial and final 'heat' distributions over a two-dimensional surface.

An explicit demonstration of the non-uniqueness of the inverse problem (on the level of f_s for a given P_{zz}) was given by Wilson and Neukirch (2011) for the case of the force-free Harris sheet, and using ideas from Schmid-Burgk (1965). As discussed in the Appendix, it is possible to rewrite the relevant integral equations in $dH_s dp_{xs} dp_{ys}$ space. When this is done, it soon becomes apparent—in the case of $\phi = 0$—that there is considerable freedom in the dependency of the DF on H_s, for a given $P_{zz}(A_x, A_y)$. This is related to the 'convoluted' nature of the (A_x, A_y) and (p_{xs}, p_{ys}) variables, and as such the $g_s(p_{xs}, p_{ys})$ function and the $P_{zz}(A_x, A_y)$ function can be considered 'tied' together, with flexibility in the function of energy.

Putting all of this together, we see that the non-uniqueness of the inverse problem can be represented by Fig. 6.1, which works as follows. For a given $\boldsymbol{B}$, one can attempt to find a self-consistent P_{zz}. In that case, one might assume the energy dependence of the DF to be of a certain form, e.g. $h = \exp(-\beta_s H_s)$, and then solve the inverse problem for g_s. Once these g_s functions are found, it may be possible to find other h functions that are self-consistent with the same P_{zz}, and hence $\boldsymbol{B}$. On top of all this, there could in practice be infinitely many such compatible P_{zz} functions (which in the force-free case can be found using established pressure transformation theory).

For each of these P_{zz} functions one could then attempt to solve the inverse problem for g_s, given an assumed form of h_s. Once this is achieved, it may be possible to generalise the energy dependency once more.

In summary, we believe that there is more work to be done regarding the non-uniqueness of Vlasov-Maxwell equilibria. It would be desirable to be able to have a 'road-map' of the variety of solutions to the inverse problem, with a clearer understanding of how they relate to one another in their mathematical structure, and their suitability for physical applications. In particular, can the somewhat complicated structure of the diagram in Fig. 6.1 be simplified, or brought in to a more holistic form, and to what extent can the heat/diffusion equation analogy be brought to bear on the problem at hand?

6.2.3 Extensions to Other Physical Systems and Geometries

Clearly, not all collisionless plasma equilibria can be modelled in a one-dimensional, Cartesian, strictly neutral and non-relativistic framework. For example, one might really need to consider two-dimensional current sheets in the Earth's magnetotail (e.g. see Artemyev and Zelenyi 2013), cylindrical geometry in a tokamak (e.g. see Tasso and Throumoulopoulos 2014), non-neutral plasmas in nonlinear electrostatic structures (e.g. see Ng et al. 2006; Vasko et al. 2016), and relativistic equilibria in pulsar magnetospheres (e.g. see DeVore et al. 2015). In contrast to the 'forward problem', the theory for the 'inverse problem' is only really well-developed for one-dimensional quaineutral plasmas in a Cartesian geometry, like those considered in this thesis. It would clearly be of interest to try and develop the methods of the inverse problem in some or all of these directions.

The generalisation that seems—at a first 'glance'—to be the most readily made, is to two-dimensional plasmas. In fact, this is the paradigm in which the 'forward problem' is most usually considered (e.g. see Schindler and Birn 2002; Schindler 2007; Artemyev and Zelenyi 2013). However, if one uses Jeans' theorem with the constants of motion of Hamiltonian and the canonical momenta, there is a clear trade-off between spatial invariance, and the number of non-zero components of the current density. To be precise, if we now let the system depend on both x and z, then p_{xs} is no longer a conserved quantity. In the absence of other conserved quantities, we now only have H_s and p_{ys} for the variables in the distribution function, and as such we can only model plasmas with a current density in the y direction, and fields that are of the form

$$\begin{aligned}
\boldsymbol{A} &= (0, A_y(x, z), 0), \\
\boldsymbol{B} &= (B_x(x, z), 0, B_z(x, z)), \\
\boldsymbol{j} &= (0, j_y(x, z), 0).
\end{aligned}$$

Note that since $j_x = j_z = 0$, we could in principle add a constant B_y field, and hence A_x, A_z that are linear functions of x, z. This would not break the self-consistency with the Vlasov approach, provided the distribution function had no dependence on A_x or A_z. This is somewhat similar to the realisation that the distribution function for the Harris sheet, is also self consistent with the Harris sheet plus guide field.

So we see there is a challenge if one wishes to maintain flexibility in both the spatial variance of the plasma considered, as well as more than one current carrying component. Formally speaking, one would have to proceed by identifying further exact (or approximate/adiabatic) constants of motion, in order to have more than one current component (e.g. see Schindler 2007; Zelenyi et al. 2011 for discussions of these topics).

The 'grand goal' of all of this theoretical work is, in my mind, some sort of unification of the forward and inverse approaches. Can we establish a framework that includes physically meaningful Vlasov-Maxwell equilibria, for which there are clear and well-understood routes from the microscopic Vlasov description of particles, to the macroscopic description of fluids and fields, and vice versa? First of all, I would be motivated to develop the forward/inverse theory—beyond quasineutrality—for distribution functions of the form described in Mottez (2004)

$$f_s(H_s, p_{xs}, p_{ys}) = \int_{a_1}^{a_2} \frac{n_{0s}(a)}{(\sqrt{2\pi}\, v_{\mathrm{th},sa})} e^{-\beta_{sa}(H_s - u_{xsa}p_{xs} - u_{ysa}p_{ys})} da,$$

for a_1, a_2 constants, and f_s the distribution function, which is formed by a continuous superposition over the index/variable a, and for which the g_s functions have been written as exponentials, i.e. eigenfunctions of the Weierstrass transform. The a variable indexes the thermal velocity, thermal beta, and the drift parameters, and f_s reduces to a more immediately recognisable distribution function when $n_{0s}(a) = \delta(a - c)n_{0s}$, for $a_1 < c < a_2$ and n_{0s} a constant. A first step in this direction might be to consider a discrete superposition rather than a continuous one, i.e. for $n_{0s}(a) = \sum_j \delta(a - a_j)$.

6.2.4 Stability

As mentioned throughout this thesis, but never really explored, a theoretical understanding of equilibria is not complete without understanding their stability properties. Knowledge of Vlasov-Maxwell equilibria allows one to study micro-instabilities in phase space (Gary 2005), for which non-thermal distribution functions are a pre-condition (i.e. multiple maxima and/or anisotropic distributions in velocity space). And keeping in mind the 'main' physical motivation for this body of work, we would be interested in considering instabilities involved in magnetic reconnection, e.g. the tearing mode (e.g. see Furth et al. 1963; Drake and Lee 1977).

There are two main approaches to assess the stability of a (kinetic) equilibrium

Normal mode analysis (e.g. see Daughton 1999; Gary 2005): Linearise the Vlasov-Maxwell equations by expressing quantities in the form $f_s = f_{0s} + f_{1s}$, $\boldsymbol{B} = \boldsymbol{B}_0 + \boldsymbol{B}_1$ etc., for the first order quantities as small perturbations to the zeroth order ones, to arrive at,

$$\frac{df_{1s}}{dt} = -\frac{q_s}{m_s}\left(\boldsymbol{E}_1 + \boldsymbol{v} \times \boldsymbol{B}_1\right) \cdot \frac{\partial f_{0s}}{\partial \boldsymbol{v}}.$$

One then subjects this equation to a Laplace/Fourier analysis in time/space (perturbed quantities $\sim e^{i(\boldsymbol{k}\cdot\boldsymbol{x}-\omega t)}$, for $\boldsymbol{k}$ the real wave-vector, and ω the complex frequency), with the aim being to solve for f_{1s}, by integrating the RHS over the 'unperturbed orbits'. One can then—in principle—use the knowledge of f_{1s} to calculate the source terms, σ_1 and $\boldsymbol{j}_1$. The source terms and the perturbed distribution function can then be substituted into the linearised Maxwell equations, from which one attempts to calculate a dispersion relation, $\omega = \omega(\boldsymbol{k})$. The results of this analysis is that for certain $\boldsymbol{k}$, and $\omega = \omega_r + i\gamma$, one should see that the equilibrium is linearly stable to some perturbations ($\gamma < 0$), and unstable to others ($\gamma > 0$). This approach does not only tell the analyst the perturbations for which the equilibrium is unstable, but it also yields the 'damping/growth-rate', $|\gamma|$, which tells us how quickly the perturbation damps/grows.

The (linear and nonlinear) energy principles: This approach counts a system as stable if "*a suitably selected test energy remains bounded by the energy supplied from external sources.*" (Schindler 2007). In the linear approach, the method essentially rests on first calculating the total energy over the spatial domain (for which there is no energy flux across the boundaries). For example, assuming the electric energy density is vanishing (consistent with quasineutrality), the energy is given by

$$W = \sum_s \int \frac{m_s}{2} v^2 f_s d^3v d^3x + \int \frac{1}{2\mu_0} B^2 d^3x.$$

Then, assuming linear perturbations of the form $f_s = f_{0s} + f_{1s}$, $\boldsymbol{B} = \boldsymbol{B}_0 + \boldsymbol{B}_1$ etc., one tries to ascertain whether—under certain dynamical constraints—there is a "*dynamic conversion of equilibrium energy into kinetic energy*" (Schindler 2007). If there is no dynamic conversion, then the equilibrium is said to be linearly stable. The energy approach typically provides sufficient criteria for stability, as opposed to necessary ones.

Preliminary analysis of the kinetic stability properties of the force-free Harris sheet have been conducted in Harrison (2009), Wilson (2013). In Wilson et al. (2016) the first particle-in-cell simulations were performed with exact initial conditions for a nonlinear force-free field. In Wilson et al. (2017) we carry out a normal-mode analysis for the collisionless tearing mode, of the manner described above, and for the Harrison-Neukirch equilibrium (Harrison and Neukirch 2009a). It is of interest to study the stability properties of exact force-free tangential equilibria—for which $\boldsymbol{B}$ ·

$\nabla = 0$ and $\nabla n = 0$—since 'density-driven/drift instabilities' (e.g. the lower hybrid drift instability) will not be present (Gary 2005).

Possible future work could include normal mode/energy principle and/or numerical (i.e. particle-in-cell) instability analyses of the specific equilibria presented in this thesis, and particularly that presented in Chap. 4, given the timely relevance to the MMS mission. One might also wish to study the stability analysis of distribution functions in a general sense, viz: "given a distribution function that is a solution of the inverse problem, what are its necessary/sufficient stability properties, and how does it grow/damp?"

References

O. Allanson, T. Neukirch, S. Troscheit, F. Wilson, From onedimensional fields to Vlasov equilibria: theory and application of Hermite polynomials. J. Plasma Phys. **82**(3), 905820306 (2016)

O. Allanson, T. Neukirch, F. Wilson, S. Troscheit, An exact collisionless equilibrium for the force-free Harris sheet with low plasma beta. Phys. Plasmas **22**(10), 102116 (2015)

O. Allanson, F. Wilson, T. Neukirch, Neutral and non-neutral collisionless plasma equilibria for twisted flux tubes: the gold-hoyle model in a background field. Phys. Plasmas **23**(9), 092106 (2016)

O. Allanson, F. Wilson, T. Neukirch, Y.-H. Liu, J.D.B. Hodgson, Exact Vlasov-Maxwell equilibria for asymmetric current sheets. Geophys. Res. Lett. **44**, 8685–8695 (2017)

W. Alpers, Steady state charge neutral models of the magnetopause. Astrophys. Space Sci. **5**, 425–437 (1969)

A. Artemyev, L. Zelenyi, Kinetic structure of current sheets in the earth magnetotail. Space Sci. Rev. **178**, 419–440 (2013)

G. Belmont, N. Aunai, R. Smets, Kinetic equilibrium for an asymmetric tangential layer. Phys. Plasmas **19**(2), 022108 (2012)

J. Birn, K. Galsgaard, M. Hesse, M. Hoshino, J. Huba, G. Lapenta, P.L. Pritchett, K. Schindler, L. Yin, J. Büchner, T. Neukirch, E.R. Priest, Forced magnetic reconnection. Geophys. Res. Lett. **32**(6), L06105 (2005)

J. Birn, M. Hesse, Energy release and transfer in guide field reconnection. Phys. Plasmas **17**(1), 012109 (2010)

J. Birn, J.F. Drake, M.A. Shay, B.N. Rogers, R.E. Denton, M. Hesse, M. Kuznetsova, Z.W. Ma, A. Bhattacharjee, A. Otto, P.L. Pritchett, Geospace environmental modeling (GEM) magnetic reconnection challenge. J. Geophys. Res. Space Phys. **106**(A3), 3715–3719 (2001)

N.A. Bobrova, S.V. Bulanov, J.I. Sakai, D. Sugiyama, Force-free equilibria and reconnection of the magnetic field lines in collisionless plasma configurations. Phys. Plasmas **8**, 759–768 (2001)

J.L. Burch, R.B. Torbert, T.D. Phan, L.-J. Chen, T.E. Moore, R.E. Ergun, J.P. Eastwood, D.J. Gershman, P.A. Cassak, M.R. Argall, S. Wang, M. Hesse, C.J. Pollock, B.L. Giles, R. Nakamura, B.H. Mauk, S.A. Fuselier, C.T. Russell, R.J. Strangeway, J.F. Drake, M.A. Shay, Y.V. Khotyaintsev, P.-A. Lindqvist, G. Marklund, F.D. Wilder, D.T. Young, K. Torkar, J. Goldstein, J.C. Dorelli, L.A. Avanov, M. Oka, D.N. Baker, A.N. Jaynes, K.A. Goodrich, I.J. Cohen, D.L. Turner, J.F. Fennell, J.B. Blake, J. Clemmons, M. Goldman, D. Newman, S.M. Petrinec, K.J. Trattner, B. Lavraud, P.H. Reiff, W. Baumjohann, W. Magnes, M. Steller, W. Lewis, Y. Saito, V. Coffey, M. Chandler, Electron-scale measurements of magnetic reconnection in space. Science **352**, aaf2939 (2016)

P.J. Channell, Exact Vlasov-Maxwell equilibria with sheared magnetic fields. Phys. Fluids **19**, 1541–1545 (1976)

W. Daughton, The unstable eigenmodes of a neutral sheet. Phys. Plasmas **6**, 1329–1343 (1999)

C.R. DeVore, S.K. Antiochos, C.E. Black, A.K. Harding, C. Kalapotharakos, D. Kazanas, A.N. Timokhin, A model for the electrically charged current sheet of a pulsar. Astrophys. J. **801**(109), 109 (2015)

N. Dorville, G. Belmont, N. Aunai, J. Dargent, L. Rezeau, Asymmetric kinetic equilibria: generalization of the BAS model for rotating magnetic profile and non-zero electric field. Phys. Plasmas **22**(9), 092904 (2015)

J.F. Drake, Y.C. Lee, Kinetic theory of tearing instabilities. Phys. Fluids **20**, 1341–1353 (1977)

L.C. Evans, Partial Differential Equations, vol. 19, 2nd edn. (Graduate Studies in Mathematics. American Mathematical Society, Providence, 2010), pp. 749

H.P. Furth, J. Killeen, M.N. Rosenbluth, Finite-resistivity instabilities of a sheet pinch. Phys. Fluids **6**, 459–484 (1963)

S.P. Gary, in *Theory of Space Plasma Microinstabilities*. Cambridge Atmospheric and Space Science Series (Cambridge University Press, 2005)

T. Gold, F. Hoyle, On the origin of solar flares. Mon. Not. Roy. Astron. Soc. **120**, 89 (1960)

E.G. Harris, On a plasma sheath separating regions of oppositely directed magnetic field. Nuovo Cimento **23**, 115 (1962)

M.G. Harrison, T. Neukirch, One-dimensional Vlasov-Maxwell equilibrium for the force-free Harris sheet. Phys. Rev. Lett. **102**(13), 135003 (2009a)

M.G. Harrison, T. Neukirch, Some remarks on one-dimensional forcefree Vlasov-Maxwell equilibria. Phys. Plasmas **16**(2), 022106 (2009b)

M.G. Harrison. Equilibrium and dynamics of collisionless current sheets. Ph.D. thesis, The University of St Andrews (2009)

M. Hesse, J. Birn, M. Kuznetsova, Collisionless magnetic reconnection: electron processes and transport modeling. J. Geophys. Res. **106**, 3721–3736 (2001)

M. Hesse, T. Neukirch, K. Schindler, M. Kuznetsova, S. Zenitani, The diffusion region in collisionless magnetic reconnection. Space Sci. Rev. **160**(1), 3–23 (2011)

M. Hesse, N. Aunai, J. Birn, P. Cassak, R.E. Denton, J.F. Drake, T. Gombosi, M. Hoshino, W. Matthaeus, D. Sibeck, S. Zenitani, Theory and modeling for the magnetospheric multiscale mission. Space Sci. Rev. **199**(1), 577–630 (2016)

A.W. Hood, P.J. Cargill, P.K. Browning, K.V. Tam, An MHD avalanche in a multi-threaded coronal loop. Astrophys. J. **817**(5), 5 (2016)

M.G. Kivelson, K.K. Khurana, Models of flux ropes embedded in a Harris neutral sheet: Force-free solutions in low and high beta plasmas. J. Geophys. Res. **100**, 23637–23646 (1995)

F. Mottez, The pressure tensor in tangential equilibria. Ann. Geophys. **22**, 3033–3037 (2004)

C.S. Ng, A. Bhattacharjee, F. Skiff, Weakly collisional Landau damping and three-dimensional Bernstein-Greene-Kruskal modes: new results on old problems. Phys. Plasmas **13**(5), 055903 (2006)

P. Ricci, J.U. Brackbill, W. Daughton, G. Lapenta, Collisionless magnetic reconnection in the presence of a guide field. Phys. Plasmas **11**, 4102–4114 (2004)

K. Schindler, *Physics of Space Plasma Activity* (Cambridge University Press, 2007)

K. Schindler, J. Birn, Models of two-dimensional embedded thin current sheets from Vlasov theory. J. Geophys. Res. (Space Phys.) **107**, 20–1 (2002)

J. Schmid-Burgk. "Zweidimensionale selbstkonsistente Lösungen der stationären Wlassowgleichung für Zweikomponentplasmen". Master's thesis, Max-Planck-Institut für Physik und Astrophysik (1965)

E. Tassi, F. Pegoraro, G. Cicogna, Solutions and symmetries of forcefree magnetic fields. Phys. Plasmas **15**(9), 092113 (2008)

H. Tasso, G. Throumoulopoulos, Tokamak-like Vlasov equilibria. Eur. Phys. J. D **68**(175), 175 (2014)

I.Y. Vasko, O.V. Agapitov, F.S. Mozer, A.V. Artemyev, J.F. Drake, Electron holes in inhomogeneous magnetic field: electron heating and electron hole evolution. Phys. Plasmas **23**(5), 052306 (2016)

A.A. Vinogradov, I.Y. Vasko, A.V. Artemyev, E.V. Yushkov, A.A. Petrukovich, L.M. Zelenyi, Kinetic models of magnetic flux ropes observed in the earth magnetosphere. Phys. Plasmas **23**(7), 072901 (2016)

T. Wiegelmann, T. Sakurai, Solar force-free magnetic fields. Living Rev. Sol. Phys. **9**(5) (2012)

F. Wilson, Equilibrium and stability properties of collisionless current sheet models. Ph.D. thesis, The University of St Andrews (2013)

F. Wilson, O. Allanson, T. Neukirch, The collisionless tearing mode in a force-free current sheet (In preparation) (2017)

F. Wilson, T. Neukirch, M. Hesse, M.G. Harrison, C.R. Stark, Particlein-cell simulations of collisionless magnetic reconnection with a nonuniform guide field. Phys. Plasmas **23**(3), 032302 (2016)

F. Wilson, T. Neukirch, A family of one-dimensional Vlasov-Maxwell equilibria for the force-free Harris sheet. Phys. Plasmas **18**(8), 082108 (2011)

L.M. Zelenyi, H.V. Malova, A.V. Artemyev, VYu. Popov, A.A. Petrukovich, Thin current sheets in collisionless plasma: equilibrium structure, plasma instabilities, and particle acceleration. Plasma Phys. Rep. **37**(2), 118–160 (2011)

Appendix A
Schmid-Burgk Variables

This Appendix is based on results in Schmid-Burgk (1965), Wilson and Neukirch (2011).

A.1 Species-Independent Integrals

For a general DF of the form $f_s = f_s(H_s, p_{xs}, p_{ys})$, we see that P_{zz} is given by

$$P_{zz} = 2\sum_s \frac{1}{m_s^3} \int_{-\infty}^{\infty}\int_{-\infty}^{\infty}\int_{H_{s,\min}}^{\infty} \sqrt{2m_s(H_s - H_{s,\min})}\, f_s dH_s dp_{xs} dp_{ys},$$

for $H_{s,\min} = [(p_{xs} - q_s A_x)^2 + (p_{ys} - q_s A_y)^2]/(2m_s) + q_s\phi$. At this stage it seems clear that the result of the integral is species-dependent. If one makes substitutions using *Schmid-Burgk variables*,

$$(E_s, P_s, Q_s) = \left(\frac{m_s H_s}{q_s^2}, \frac{p_{xs}}{q_s}, \frac{p_{ys}}{q_s}\right),$$

$$F_s(E_s, P_s, Q_s) = \frac{m_s^3}{q_s^4} f_s(H_s, p_{xs}, p_{ys}),$$

then P_{zz} is now written

$$P_{zz} = 2\sum_s \frac{e}{m_s} \int_{-\infty}^{\infty}\int_{-\infty}^{\infty}\int_{E_{s,\min}}^{\infty} \sqrt{(E_s - E_{s,\min})}\, F_s dE_s dP_s dQ_s,$$

for $E_{s,\min} = [(P_s - A_x)^2 + (Q_s - A_y)^2]/2 + \frac{q_s}{m_s}\phi$. As yet, we have only made substitutions, and there have been no restrictions. However, if we now assume strict neutrality, $\phi = 0$, and—crucially—assume that the *functional form* of the F_s function is independent of species, then the above expression has an interesting property.

O. Allanson, *Theory of One-Dimensional Vlasov-Maxwell Equilibria*, Springer Theses, https://doi.org/10.1007/978-3-319-97541-2

Note that when we say 'functional form is independent of species', we mean that regardless of the species s, the function F_s maps the inputs (E_s, P_s, Q_s) according to the same rules, i.e.

$$F_s(E_s, P_s, Q_s) = F(E_s, P_s, Q_s),$$

(for example, it cannot use an exponential function for ions, and a quadratic function for electrons). Under these assumptions, the triple integral in the P_{zz} expression actually becomes species-independent. The (E_s, P_s, Q_s) variables are nothing but dummy variables, and the integrand itself is now of the same form, regardless of s. As a result, P_{zz} becomes

$$P_{zz}(A_x, A_y) = 2e\left(\frac{1}{m_e} + \frac{1}{m_i}\right)\int_{-\infty}^{\infty}\int_{-\infty}^{\infty}\int_{E_{s,\min}}^{\infty}\sqrt{(E_s - E_{s,\min})}F dE_s dP_s dQ_s. \tag{A.1}$$

Similarly it can be shown that the charge density is given by

$$\sigma(A_x, A_y) = 2\int_{-\infty}^{\infty}\int_{-\infty}^{\infty}\int_{E_{s,\min}}^{\infty}(E_s - E_{s,\min})^{-1/2}F dE_s dP_s dQ_s \sum_s \frac{q_s}{e} = 0,$$

and we see that the DF is automatically self-consistent with the assumption of strict neutrality.

The Schmid-Burgk variables have helped us to demonstrate that the species-dependency of velocity moments of the DF enter through a q_s/m_s factor that multiplies the scalar potential, and through any 'innate' species-dependency that the DFs may have in themselves. In particular, the assumption of strict neutrality is automatically self-consistent if $F_s = F$ (in the case of an electron-ion plasma, or any plasma for which $\sum_s q_s/|q_s| = 0$).

A.1.1 Freedom in the Energy Dependency

Using the Schmid-Burgk variables and the assumptions explained above ($\phi = 0$, $F_s = F$), Wilson and Neukirch (2011) show—for the the example of the FFHS—that it is possible under certain conditions to solve the inverse problem with a DF of the general form

$$F = h(E_s)g(p_{xs}, p_{ys}),$$

and with the h function not only of the typically assumed exponential form, but of a reasonably arbitrary nature. This process is demonstrated for h functions that are in Dirac delta form ($\delta(E_s - E_0)$), Step function form ($\Theta(E_0 - E_s)$), and polynomial form ($\Theta(E_0 - E_s)\,(E_0 - E_s)^{\chi}$, for $\chi > -1$).

As such, we can consider the $P_{zz}(A_x, A_y)$ and $g_s(p_{xs}, p_{ys})$ functions as 'tied' together. This 'tie' is evidenced by the convoluted nature of the variables $\boldsymbol{A}$ and $\boldsymbol{p}_s$ in the relevant integral equations, i.e. velocity moments of the DF, in general form, are given by

$$\langle v_j^k f_s\rangle(A_x, A_y) := \frac{n_{0s}}{(\sqrt{2\pi}\,v_{\mathrm{th},s})^3}\frac{2}{m_s^{k+2}}\times$$
$$\int_{-\infty}^{\infty}\int_{-\infty}^{\infty}\int_{H_{s,\min}}^{\infty}\frac{(p_{js}-q_sA_j)^k}{\sqrt{2m_s(H_{s,\min}-H_s)}}f_s(H_s, p_{xs}, p_{ys})dH_sdp_{xs}dp_{ys}.$$

A.1.2 Summary

In summary, the Schmid-Burgk variables have helped us to see that in the case of strictly neutral plasmas, there is evidence to suggest that the inverse problem should be framed as as: *"for a given macroscopic equilibrium, i.e. a $P_{zz}(a_x, A_y)$, what are the self-consistent g functions"*, for

$$f_s \propto h(E_s)g(P_s, Q_s),$$

as opposed to: *"for a given macroscopic equilibrium, i.e. a $P_{zz}(a_x, A_y)$, what are the self-consistent DFs?"*

References

J. Schmid-Burgk, Zweidimensionale selbstkonsistente Lösungen der stationären Wlassowgleichung für Zweikomponentplasmen (1965). Max-Planck-Institut für Physik und Astrophysik, Master's thesis

F. Wilson, T. Neukirch, A family of one-dimensional Vlasov-Maxwell equilibria for the force-free Harris sheet. Phys. Plasmas **18**(8), 082108 (2011)

Printed in the United States
By Bookmasters